T0296835

Modeling of Post-Myocardial Infarction

Modeling of Post-Myocardial Infarction

ODE/PDE Analysis with R

William E. Schiesser

Lehigh University
Bethlehem, PA, United States

ELSEVIER

ACADEMIC PRESS
An imprint of Elsevier

Academic Press is an imprint of Elsevier
125 London Wall, London EC2Y 5AS, United Kingdom
525 B Street, Suite 1650, San Diego, CA 92101, United States
50 Hampshire Street, 5th Floor, Cambridge, MA 02139, United States
The Boulevard, Langford Lane, Kidlington, Oxford OX5 1GB, United Kingdom

Notices

Knowledge and best practice in this field are constantly changing. As new research and experience broaden our understanding, changes in research methods, professional practices, or medical treatment may become necessary.

Practitioners and researchers must always rely on their own experience and knowledge in evaluating and using any information, methods, compounds, or experiments described herein. In using such information or methods they should be mindful of their own safety and the safety of others, including parties for whom they have a professional responsibility.

To the fullest extent of the law, neither the Publisher nor the authors, contributors, or editors, assume any liability for any injury and/or damage to persons or property as a matter of products liability, negligence or otherwise, or from any use or operation of any methods, products, instructions, or ideas contained in the material herein.

ISBN: 978-0-443-13611-5

For information on all Academic Press publications
visit our website at https://www.elsevier.com/books-and-journals

Publisher: Mara E. Conner
Acquisitions Editor: Chris Katsaropoulos
Editorial Project Manager: Howi M. De Ramos
Production Project Manager: Erragounta Saibabu Rao
Cover Designer: Miles Hitchen

Typeset by VTeX

Contents

Preface

Survivors of a myocardial-infaction (MI) are likely to experience a reduced (impaired) cardiac function. This results from a series of post-MI biomolecular reactions that are modeled in this book by systems of ordinary and partial differential equations (ODE/PDEs) [1–3].

Initially, monocytes and myocytes are produced in a post-MI that then react to produce macrophages and cytokines, that may adversely affect the cardiac tissue such as inflammation and reduction (weakening) of the extra cellular matrix (ECM).

The first model developed and analyzed by computer-based numerical methods is a system of six ODEs with time as the independent variable and the following dependent variables:

variable	interpretation in [3]
M_{un}	cell density of unactivated macrophage
M_1	cell density of M_1 macrophage
M_2	cell density of M_2 macrophage
IL_{10}	concentration of IL_{10} (interleuken-10)
T_α	concentration of $TNF - \alpha$ (tumor necrosis factor-α)
IL_1	concentration of IL_1 (interleuken-1)

The system of six ODEs does not include a spatial aspect of a MI in the cardiac tissue. Therefore, the ODE model is extended to include a spatial effect by the addition of diffusion terms. The resulting system of six diffusion PDEs, with x (space) and t (time) as independent variables, is integrated (solved) by the numerical method of lines (MOL), a general numerical algorithm for PDEs.

The infarction of the MI can be spatially variable within the PDE diffusion (cardiac tissue) model. This spatial variation is analyzed through changes of the monocyte and myocyte volumetric generation rates.

The properties of the six PDE model with dependent variables

$$M_{un}(x,t), \ M_1(x,t), \ M_2(x,t), \ IL_{10}(x,t), \ T_\alpha(x,t), \ IL_1(x,t)$$

are elucidated by computation and display of the PDE RHS terms and LHS derivatives in t,

$$\frac{\partial M_{un}(x,t)}{\partial t}, \ \frac{\partial M_1(x,t)}{\partial t}, \ \frac{\partial M_2(x,t)}{\partial t}, \ \frac{\partial IL_{10}(x,t)}{\partial t}, \ \frac{\partial T_\alpha(x,t)}{\partial t}, \ \frac{\partial IL_1(x,t)}{\partial t}$$

To conclude the book, an analysis of the right hand side terms of the six PDE model is implemented which gives a detailed explanation of the PDE t derivatives and numerical solutions.

References

[1] K. Soetaert, J. Cash, F. Mazzia, Solving Differential Equations in R, Springer-Verlag, Heidelberg, Germany, 2012.

[2] O.F. Voropaeva, Ch.E. Tsgoev, A numerical model of inflammation dynamics in the core of muocardial infarction, Journal of Applied and Industrial Mathematics 13 (2) (2019) 372–383.

[3] Y. Wang, et al., Mathematical modeling and stability analysis of macrophage activation in left ventricular remodeling post-myocardial infarction, BMC Genomics 13 (Suppl 6) (2012).

ODE model development

Introduction

Myocytes and monocytes from a post-myocardial infarction (post-MI) react to produce macrophages and cytokines that are harmful to cardiac tissue, e.g., inflammation and reduction (weakening) of the extra cellular matrix (ECM). The biomolecular reactions are discussed in Chapter 1, including a mathematical model of six ordinary differential equations (ODEs).

1.1 Formulation of ODE post-MI model

The initial ODE model is stated as eqs. (1.1) [3,4].

$$\frac{dM_{un}}{dt} = M - k_2 M_{un} \frac{IL_1}{IL_1 + c_{IL_1}} - k_3 M_{un} \frac{T_\alpha}{T_\alpha + c_{T_\alpha}}$$
$$- k_4 M_{un} \frac{IL_{10}}{IL_{10} + c_{IL_{10}}} - \mu M_{un} \qquad (1.1\text{-}1)$$

$$\frac{dM_1}{dt} = k_2 M_{un} \frac{IL_1}{IL_1 + c_{IL_1}} + k_3 M_{un} \frac{T_\alpha}{T_\alpha + c_{T_\alpha}}$$
$$+ k_1' M_2 - k_1 M_1 - \mu M_1 \qquad (1.1\text{-}2)$$

$$\frac{dM_2}{dt} = k_4 M_{un} \frac{IL_{10}}{IL_{10} + c_{IL_{10}}}$$
$$+ k_1 M_1 - k_1' M_2 - \mu M_2 \qquad (1.1\text{-}3)$$

$$\frac{dIL_{10}}{dt} = k_5 M_2 \frac{c_1}{c_1 + IL_{10}} - d_{IL_{10}} IL_{10} \qquad (1.1\text{-}4)$$

$$\frac{dT_\alpha}{dt} = (k_6 M_1 + \lambda M_c) \frac{c}{c + IL_{10}} - d_{T_\alpha} T_\alpha \qquad (1.1\text{-}5)$$

$$\frac{dIL_1}{dt} = (k_7 M_1 + \lambda M_c) \frac{c}{c + IL_{10}} - d_{IL_1} IL_1 \qquad (1.1\text{-}6)$$

The six dependent variables and one independent variable are explained in Table 1.1.

Modeling of Post-Myocardial Infarction
https://doi.org/10.1016/B978-0-44-313611-5.00006-8

Table 1.1: Dependent, independent variables of eqs. (1.1)

variable	interpretation in [3,4]
M_{un}	cell density of unactivated macrophage
M_1	cell density of M_1 macrophage
M_2	cell density of M_2 macrophage
IL_{10}	concentration of IL_{10} (interleuken-10)
T_α	concentration of $TNF - \alpha$ (tumor necrosis factor-α)
IL_1	concentration of IL_1 (interleuken-1)
t	time

In summary, monocytes through the term M, and myocytes through the term λM_c produce three macrophages with densities $M_{nu}(t)$, $M_1(t)$, $M_2(t)$ and three cytokines with concentrations IL_{10}, $T_\alpha(t)$, $IL_1(t)$ as the solutions of eqs. (1.1), (1.2).

Eq. (1.1-1) is a dynamic balance on the unactivated macrophage. The terms in eq. (1.1-1) are time rates as explained next.

Table 1.2-1: Explanation of terms in eq. (1.1-1)

1. $\dfrac{dM_{un}}{dt}$: time rate (derivative) of M_{un}.
2. M: differentiation of monocyte to inactivated macrophage.
3. $-k_2 M_{un} \dfrac{IL_1}{IL_1 + c_{IL_1}}$: M_{un} activated to M_1 via IL_1. This term is a nonlinear coupling between eqs. (1.1-1) and (1.1-6).
4. $-k_3 M_{un} \dfrac{T_\alpha}{T_\alpha + c_{T_\alpha}}$: M_{un} activated to M_1 via T_α.
5. $-k_4 M_{un} \dfrac{IL_{10}}{IL_{10} + c_{IL_{10}}}$: M_{un} activated to M_2 via IL_{10}.
6. $-\mu M_{un}$: M_{un} depletion.

Eq. (1.1-2) is a dynamic balance on the M_1 macrophage. The terms in eq. (1.1-2) are time rates as explained next.

Table 1.2-2: Explanation of terms in eq. (1.1-2)

1. $\dfrac{dM_1}{dt}$: time rate (derivative) of M_1.

2. $k_2 M_{un} \dfrac{IL_1}{IL_1 + c_{IL_1}}$: M_{un} activated to M_1 via IL_1.

3. $k_3 M_{un} \dfrac{T_\alpha}{T_\alpha + c_{T_\alpha}}$: M_{un} activated to M_1 via T_α.

4. $k'_1 M_2 - k_1 M_1$: M_1 - M_2 exchange.

5. $-\mu M_1$: M_1 depletion.

Eq. (1.1-3) is a dynamic balance on the M_2 macrophage. The terms in eq. (1.1-3) are time rates as explained next.

Table 1.2-3: Explanation of terms in eq. (1.1-3)

1. $\dfrac{dM_2}{dt}$: time rate (derivative) of M_2.

2. $k_4 M_{un} \dfrac{IL_{10}}{IL_{10} + c_{IL_{10}}}$: M_{un} activated to M_2 via IL_{10}.

3. $k_1 M_1 - k'_1 M_2-$: M_1 - M_2 exchange.

4. $-\mu M_2$: M_2 depletion.

Eq. (1.1-4) is a dynamic balance on IL_{10}. The terms in eq. (1.1-4) are time rates as explained next.

Table 1.2-4: Explanation of terms in eq. (1.1-4)

1. $\dfrac{dIL_{10}}{dt}$: time rate (derivative) of IL_{10}.

2. $k_5 M_2 \dfrac{c_1}{c_1 + IL_{10}}$: IL_{10} secreted by M_2 via IL_{10}.

3. $-\mu IL_{10}$: IL_{10} depletion.

Eq. (1.1-5) is a dynamic balance on the $TNF - \alpha$. The terms in eq. (1.1-5) are time rates as explained next.

Table 1.2-5: Explanation of terms in eq. (1.1-5)

1. $\dfrac{dT_\alpha}{dt}$: time rate (derivative) of T_α.

2. $(k_6 M_1 + \lambda M_c)\dfrac{c}{c + IL_{10}}$: T_α secreted by M_1
    $(k_6 M_1)$ and myocytes (λM_c) via IL_{10}.

3. $-d_{T_\alpha} T_\alpha$: T_α depletion.

Eq. (1.1-6) is a dynamic balance on the IL_1. The terms in eq. (1.1-6) are time rates as explained next.

Table 1.2-6: Explanation of terms in eq. (1.1-6)

1. $\dfrac{dIL_1}{dt}$: time rate (derivative) of IL_1.

2. $(k_7 M_1 + \lambda M_c)\dfrac{c}{c + IL_{10}}$: IL_1 secreted by M_1
    $(k_7 M_1)$ and myocytes (λM_c) via IL_{10}.

3. $-d_{IL_1} IL_1$: $T_I L_1$ depletion.

Eqs. (1.1-1,2,3) pertain to the biomolecular production rate of three macrophages, M_{un}, M_1, M_2. Eqs. (1.1-4,5,6) pertain to the biomolecular production rate of three cytokines, IL_{10}, $TNF - \alpha$, TL_1. The coupling between eqs. (1.1) is illustrated by a diagram in [4], Fig. 1. The parameters (constants) in eqs. (1.1) are discussed in Chapter 2.

Eqs. (1.1) are first order in t and therefore each requires one initial condition (IC).

$$M_{un}(t = 0) = M_{un,0} \qquad (1.2\text{-}1)$$

$$M_1(t = 0) = M_{1,0} \qquad (1.2\text{-}2)$$

$$M_2(t = 0) = M_{2,0} \qquad (1.2\text{-}3)$$

$$IL_{10}(t = 0) = IL_{10,0} \qquad (1.2\text{-}4)$$

$$T_\alpha(t = 0) = T_{\alpha,0} \qquad (1.2\text{-}5)$$

$$IL_1(t = 0) = IL_{1,0} \qquad (1.2\text{-}6)$$

where $M_{un,0}$ to $IL_{1,0}$ are constants to be specified.

Eqs. (1.1), (1.2) constitute the initial ODE post-MI model implemented in Chapter 2 as a system of R routines[1].

---

[1]   R is a quality, open-source scientific computing system readily downloaded from the Internet.

1.2 Summary and conclusions

Eqs. (1.1) with ICs (1.2) constitute a basic ODE model for the analysis of post-MI biomolecular reactions. This ODE system is implemented in Chapter 2 as a set of R routines [2]. Additional background pertaining to post-MI is given in [1].

References

[1] M.C. Bahit, A. Kochar, C.B. Granger, Post-myocardial infarction heart failure, Journal of the American College of Cardiology 6 (3) (2018) 179–186.

[2] K. Soetaert, J. Cash, F. Mazzia, Solving Differential Equations in R, Springer-Verlag, Heidelberg, Germany, 2012.

[3] O.F. Voropaeva, Ch.E. Tsgoev, A numerical model of inflammation dynamics in the core of myocardial infarction, Journal of Applied and Industrial Mathematics 13 (2) (2019) 372–383.

[4] Y. Wang, et al., Mathematical modeling and stability analysis of macrophage activation in left ventricular remodeling post-myocardial infarction, BMC Genomics 13 (Suppl 6) (2012).

ODE model implementation

## CHAPTER 2

Introduction

The computer implementation of the post-MI model of eqs. (1.1), (1.2) gives $M_{nu}(t)$, $M_1(t)$, $M_2(t)$, $IL_{10}(t)$, $T_\alpha(t)$, $IL_1(t)$ (Table 1.1). The solution is displayed numerically and graphically by the main program listed next.

2.1 Coding of the post-MI model

The following main program for eqs. (1.1), (1.2) is based on the use of the basic R system [2].

2.1.1 Main program

```
#
# Six ODE post-MI model
#
# Delete previous workspaces
  rm(list=ls(all=TRUE))
#
# Access ODE integrator
  library("deSolve");
#
# Access functions for numerical solution
  setwd("g:/myocardial infarction/chap2");
  source("ode1a.R");
#
# Parameters
  Mun_0=0;
    M1_0=0;
    M2_0=0;
  IL10_0=0;
    Ta_0=0;
   IL1_0=0;
      k1=1;
```

Modeling of Post-Myocardial Infarction
https://doi.org/10.1016/B978-0-44-313611-5.00007-X

```
    k1p=1;
     k2=1;
     k3=1;
     k4=1;
     k5=1;
     k6=1;
     k7=1;
      c=1;
     c1=1;
   cIL1=1;
    cTa=1;
  cIL10=1;
  dIL10=1;
    dTa=1;
   dIL1=1;
      M=0;
    lam=1;
     Mc=0;
     mu=1;
#
# Independent variable for ODE integration
  t0=0;tf=1;nout=11;
  tout=seq(from=t0,to=tf,by=(tf-t0)/(nout-1));
#
# Initial conditions
  u0=rep(0,6);
  u0[1]=Mun_0;
  u0[2]=M1_0;
  u0[3]=M2_0;
  u0[4]=IL10_0;
  u0[5]=Ta_0;
  u0[6]=IL1_0;
  ncall=0;
#
# ODE integration
  out=lsodes(y=u0,times=tout,func=ode1a,
      sparsetype ="sparseint",rtol=1e-6,
      atol=1e-6,maxord=5);
  nrow(out)
```

```
  ncol(out)
#
# Arrays for plotting numerical solution
  Munp=rep(0,nout);
   M1p=rep(0,nout);
   M2p=rep(0,nout);
  IL10p=rep(0,nout);
   Tap=rep(0,nout);
  IL1p=rep(0,nout);
  for(it in 1:nout){
    Munp[it]=out[it,2];
     M1p[it]=out[it,3];
     M2p[it]=out[it,4];
   IL10p[it]=out[it,5];
     Tap[it]=out[it,6];
    IL1p[it]=out[it,7];
  }
#
# Plot ODE solutions
#
# Mun
  par(mfrow=c(2,2));
  matplot(x=tout,y=Munp,type="l",xlab="t",ylab="Mun(t)",
          xlim=c(t0,tf),lty=1,main="Mun(t)",lwd=2,
          col="black");#
# M1
  matplot(x=tout,y=M1p,type="l",xlab="t",ylab="M1(t)",
          xlim=c(t0,tf),lty=1,main="M1(t)",lwd=2,
          col="black");
#
# M2
  matplot(x=tout,y=M2p,type="l",xlab="t",ylab="M2(t)",
          xlim=c(t0,tf),lty=1,main="M2(t)",lwd=2,
          col="black");
#
# IL10
  matplot(x=tout,y=IL10p,type="l",xlab="t",ylab="IL10(t)",
          xlim=c(t0,tf),lty=1,main="IL10(t)",lwd=2,
          col="black");
```

```
#
# Ta
  matplot(x=tout,y=Tap,type="l",xlab="t",ylab="Ta(t)",
          xlim=c(t0,tf),lty=1,main="Ta(t)",lwd=2,
          col="black");
#
# IL1

  matplot(x=tout,y=IL1p,type="l",xlab="t",ylab="IL1(t)",
          xlim=c(t0,tf),lty=1,main="Il1(t)",lwd=2,
          col="black");
```

Listing 2.1: Main program for eqs. (1.1), (1.2)

We can note the following details about Listing 2.1.

- Previous workspaces are deleted.

```
#
# Six ODE post-MI model
#
# Delete previous workspaces
  rm(list=ls(all=TRUE))
```

- The R ODE integrator library deSolve is accessed [2]. Then the directory with the files for the solution of eqs. (1.1), (1.2) is designated. Note that setwd (set working directory) uses / rather than the usual \.

```
#
# Access ODE integrator
  library("deSolve");
#
# Access functions for numerical solution
  setwd("f:/myocardial infarction/chap2");
  source("ode1a.R");
```

ode1a is the routine for eqs. (1.1) discussed subsequently.
- The model parameters are specified numerically.

```
#
# Parameters
    Mun_0=0;
```

```
    M1_0=0;
    M2_0=0;
  IL10_0=0;
    Ta_0=0;
   IL1_0=0;
  k1=1;
 k1p=1;
  k2=1;
  k3=1;
  k4=1;
  k5=1;
  k6=1;
  k7=1;
     c=1;
    c1=1;
  cIL1=1;
   cTa=1;
 cIL10=1;
 dIL10=1;
   dTa=1;
  dIL1=1;
  M=0;
 lam=1;
  Mc=0;
  mu=1;
```

The parameters are named according to eqs. (1.1), (1.2), e.g.,

```
  Mun_0=0;
    M1_0=0;
    M2_0=0;
  IL10_0=0;
    Ta_0=0;
   IL1_0=0;
```

are the ICs of eqs. (1.2).

* An interval in t is defined for 11 output points, so that tout=0,0.1,...,1.

```
#
# Independent variable for ODE integration
  t0=0;tf=1;nout=11;
  tout=seq(from=t0,to=tf,by=(tf-t0)/(nout-1));
```

- ICs (1.2) are programmed.

```
#
# Initial conditions
  u0=rep(0,6);
  u0[1]=Mun_0;
  u0[2]=M1_0;
  u0[3]=M2_0;
  u0[4]=IL10_0;
  u0[5]=Ta_0;
  u0[6]=IL1_0;
  ncall=0;
```

Also, the counter for the calls to ode1a is initialized.

- The system of 6 ODEs is integrated by the library integrator lsodes (available in deSolve, [2]). As expected, the inputs to lsodes are the ODE function, ode1a, the IC vector u0, and the vector of output values of t, tout. The length of u0 (6) informs lsodes how many ODEs are to be integrated. func,y,times are reserved names.

```
#
# ODE integration
  out=lsodes(y=u0,times=tout,func=ode1a,
      sparsetype ="sparseint",rtol=1e-6,
      atol=1e-6,maxord=5);
  nrow(out)
  ncol(out)
```

nrow,ncol confirm the dimensions of out.

- $M_{nu}(t)$, $M_1(t)$, $M_2(t)$, $IL_{10}(t)$, $T_\alpha(t)$, $IL_1(t)$ are placed in arrays for numerical and graphical output.

```
#
# Arrays for plotting numerical solution
   Munp=rep(0,nout);
    M1p=rep(0,nout);
    M2p=rep(0,nout);
  IL10p=rep(0,nout);
    Tap=rep(0,nout);
   IL1p=rep(0,nout);
  for(it in 1:nout){
```

```
   Munp[it]=out[it,2];
    M1p[it]=out[it,3];
    M2p[it]=out[it,4];
  IL10p[it]=out[it,5];
    Tap[it]=out[it,6];
   IL1p[it]=out[it,7];
  }
```

The offset +1 is required because the first element of the solution vectors in out is the value of t and the 2 to 7 elements are the 6 values of $M_{nu}(t)$, $M_1(t)$, $M_2(t)$, $IL_{10}(t)$, $T_\alpha(t)$, $IL_1(t)$. These dimensions from the preceding calls to nrow,ncol are confirmed in the subsequent output.

• $M_{nu}(t)$, $M_1(t)$, $M_2(t)$, $IL_{10}(t)$, $T_\alpha(t)$, $IL_1(t)$ are plotted with the matplot utility. par(mfrow=c(2,2)) specifies a 2×2 matrix of plots on a page of graphical output.

```
#
# Plot ODE solutions
#
# Mun
  par(mfrow=c(2,2));
  matplot(x=tout,y=Munp,type="l",xlab="t",ylab="Mun(t)",
          xlim=c(t0,tf),lty=1,main="Mun(t)",lwd=2,
          col="black");
#
# M1
  matplot(x=tout,y=M1p,type="l",xlab="t",ylab="M1(t)",
          xlim=c(t0,tf),lty=1,main="M1(t)",lwd=2,
          col="black");
#
# M2
  matplot(x=tout,y=M2p,type="l",xlab="t",ylab="M2(t)",
          xlim=c(t0,tf),lty=1,main="M2(t)",lwd=2,
          col="black");
#
# IL10
  matplot(x=tout,y=IL10p,type="l",xlab="t",ylab="IL10(t)",
          xlim=c(t0,tf),lty=1,main="IL10(t)",lwd=2,
          col="black");
#
```

```
  # Ta
    matplot(x=tout,y=Tap,type="l",xlab="t",ylab="Ta(t)",
            xlim=c(t0,tf),lty=1,main="Ta(t)",lwd=2,
            col="black");
  #
  # IL1

    matplot(x=tout,y=IL1p,type="l",xlab="t",ylab="IL1(t)",
            xlim=c(t0,tf),lty=1,main="Tl1(t)",lwd=2,
            col="black");
```

This completes the discussion of the main program for eqs. (1.1), (1.2). The ODE routine
ode1a called by lsodes from the main program (Listing 2.1) for the numerical integration
of eqs. (1.1), (1.2) is next.

2.1.2 ODE routine

ode1a called in the main program of Listing 2.1 follows.

```
  ode1a=function(t,u,parms){
#
# Function ode1a computes the t derivative
# vectors of Mun(t),M1(t),M2(t),IL10(t),
# Talpha(t),IL1(t)
#
# One vector to six scalars
   Mun=u[1];
    M1=u[2];
    M2=u[3];
  IL10=u[4];
    Ta=u[5];
   IL1=u[6];
#
# ODEs
   fIL1=k2*Mun*IL1/(IL1+cIL1);
    fTa=k3*Mun*Ta/(Ta+cTa);
  fIL10=k4*Mun*IL10/(IL10+cIL10);
    fc1=k5*M2*c1/(c1+IL10);
     fc=c/(c+IL10);
   Munt=M-fIL1-fTa-fIL10-mu*Mun;
```

```
   M1t=fIL1+fTa+k1p*M2-k1*M1-mu*M1;
   M2t=fIL10+k1*M1-k1p*M2-mu*M2;
 IL10t=fc1-dIL10*IL10;
   Tat=(k6*M1+lam*Mc)*fc-dTa*Ta;
  IL1t=(k7*M1+lam*Mc)*fc-dIL1*IL1;
#
# Six scalars to one vector
  ut=rep(0,6);
  ut[1]=Munt;
  ut[2]=M1t;
  ut[3]=M2t;
  ut[4]=IL10t;
  ut[5]=Tat;
  ut[6]=IL1t;
#
# Increment calls to ode1a
  ncall <<- ncall+1;
#
# Return derivative vector
  return(list(c(ut)));
  }
```

Listing 2.2: ODE routine ode1a for eqs. (1.1)

We can note the following details about ode1a.

- The function is defined.

  ```
    ode1a=function(t,u,parms){
  #
  # Function ode1a computes the t derivative
  # vectors of Mun(t),M1(t),M2(t),IL10(t),
  # Talpha(t),IL1(t)
  ```

  t is the current value of *t* in eqs. (1.1). u is the 6-vector of ODE dependent variables.
  parm is an argument to pass parameters to ode1a (unused, but required in the argument
  list). The arguments must be listed in the order stated to properly interface with lsodes
  called in the main program of Listing 2.1. The derivative vector of the LHS of eqs. (1.1) is
  calculated and returned to lsodes as explained subsequently.
- Vector u is placed in six scalars to facilitate the programming of eqs. (1.1).

```
#
# One vector to six scalars
   Mun=u[1];
    M1=u[2];
    M2=u[3];
  IL10=u[4];
    Ta=u[5];
   IL1=u[6];
```

- Eqs. (1.1) are programmed, starting with the nonlinear rate functions.

```
#
# ODEs
   fIL1=k2*Mun*IL1/(IL1+cIL1);
    fTa=k3*Mun*Ta/(Ta+cTa);
 fIL10=k4*Mun*IL10/(IL10+cIL10);
   fc1=k5*M2*c1/(c1+IL10);
     fc=c/(c+IL10);
```

The coding of the nonlinear functions is explained briefly in the following table.

1.	`fIL1=k2*Mun*IL1/(IL1+cIL1)`: $k_2 M_{un} \dfrac{IL_1}{IL_1 + c_{IL_1}}$ in eqs. (1.1-1,2).
2.	`fTa=k3*Mun*Ta/(Ta+cTa)`: $k_3 M_{un} \dfrac{T_\alpha}{T_\alpha + c_{T_\alpha}}$ in eqs. (1.1-1,2).
3.	`fIL10=k4*Mun*IL10/(IL10+cIL10)`: $k_4 M_{un} \dfrac{IL_{10}}{IL_{10} + c_{IL_{10}}}$ in eqs. (1.1-1,3).
4.	`fc1=k5*M2*c1/(c1+IL10)`: $k_5 M_2 \dfrac{c_1}{c_1 + IL_{10}}$ in eq. (1.1-4).
5.	`fc=c/(c+IL10)`: $\dfrac{c}{c + IL_{10}}$ in eqs. (1.1-5,6).

These examples demonstrate the straightforward use of nonlinear functions numerically, which would be difficult to accomplish analytically. Also, since these functions are computed in the programming of the ODEs, they can be displayed graphically (plotted) as well as the ODE dependent variables, which gives a direct indication of the contribution of the nonlinear functions.

- Eqs. (1.1) are programmed.

```
Munt=M-fIL1-fTa-fIL10-mu*Mun;
M1t=fIL1+fTa+k1p*M2-k1*M1-mu*M1;
M2t=fIL10+k1*M1-k1p*M2-mu*M2;
```

```
IL10t=fc1-dIL10*IL10;
  Tat=(k6*M1+lam*Mc)*fc-dTa*Ta;
  IL1t=(k7*M1+lam*Mc)*fc-dIL1*IL1;
```

For example, the left hand side (LHS) derivative $\dfrac{\partial M_{un}}{\partial t}$ in eq. (1.1-1) is programmed as

Munt, and the RHS $M - k_2 M_{un} \dfrac{IL_1}{IL_1 + c_{IL_1}} - k_3 M_{un} \dfrac{T_\alpha}{T_\alpha + c_{T_\alpha}} - k_4 M_{un} \dfrac{IL_{10}}{IL_{10} + c_{IL_{10}}} -$

μM_{un} is programmed as M-fIL1-fTa-fIL10-mu*Mun.

- The 6 ODE derivatives are placed in the vector ut for return to lsodes to take the next step in *t* along the solution.

```
#
# Six scalars to one vector
  ut=rep(0,6);
  ut[1]=Munt;
  ut[2]=M1t;
  ut[3]=M2t;
  ut[4]=IL10t;
  ut[5]=Tat;
  ut[6]=IL1t;
```

- The counter for the calls to pde1a is incremented and returned to the main program of Listing 2.1 by <<-.

```
#
# Increment calls to ode1a
  ncall <<- ncall+1;
```

- The vector ut is returned as a list as required by lsodes. c is the R vector utility. The final } concludes ode1a.

```
#
# Return derivative vector
  return(list(c(ut)));
  }
```

This completes the discussion of ode1a. The output from the main program of Listing 2.1 and ODE routine ode1a of Listing 2.2 is considered next.

2.1.3 Numerical, graphical output

The output from nrow,ncol called in the main program of Listing 2.1 (dimensions of the solution matrix out from lsodes) follows.

[1] 11

[1] 7

The row dimension of the solution matrix out from lsodes is 11 corresponding to the definition of the t values of tout.

```
#
# Independent variable for ODE integration
  t0=0;tf=1;nout=11;
  tout=seq(from=t0,to=tf,by=(tf-t0)/(nout-1));
```

The column dimension of out is 7 corresponding to the 6 ODE dependent variables $M_{un}(t)$ to $IL_1(t)$ of eqs. (1.1) and the independent variable t for the 6 dependent variables.

The graphical output of Figs. 2.1 indicates that the six dependent variables remain at the homogeneous ICs of eqs. (1.2).

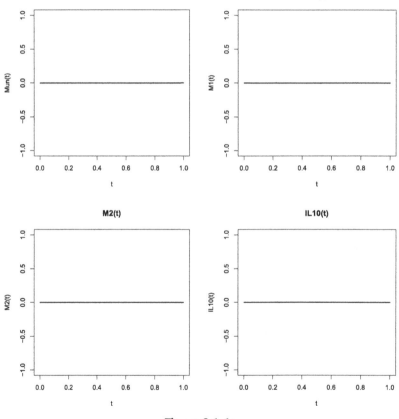

Figure 2.1-1:
Numerical $M_{nu}(t)$, $M_1(t)$, $M_2(t)$, $IL_{10}(t)$ from eqs. (1.1), (1.2)

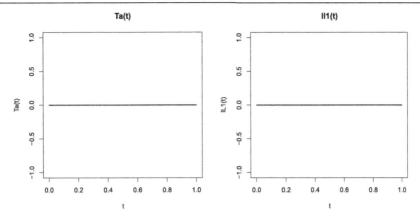

Figure 2.1-2:
Numerical $T_\alpha(t)$, $IL_1(t)$ from eqs. (1.1), (1.2)

The six ODE dependent variables remain at the homogeneous (zero) ICs since all of the RHS terms of the ODEs are zero. This is an important special case since a departure from the zero ICs would indicate a programming error.

As an example of an ODE dependent variable that remains at the IC, $M_{un}(t)$ from eq. (1.1-1) is analyzed in Table 2.1 for zero ICs.

Table 2.1: Explanation of RHS terms in eq. (1.1-1) for zero ICs

1. $M = 0$: Set as a parameter in the main program of Listing 2.1.

2. $-k_2 M_{un} \dfrac{IL_1}{IL_1 + c_{IL_1}} = -k_2 0 \dfrac{0}{0 + c_{IL_1}} = 0$

3. $-k_3 M_{un} \dfrac{T_\alpha}{T_\alpha + c_{T_\alpha}} = -k_3 0 \dfrac{0}{0 + c_{T_\alpha}} = 0$

4. $-k_4 M_{un} \dfrac{IL_{10}}{IL_{10} + c_{IL_{10}}} = -k_4 0 \dfrac{0}{0 + c_{IL_{10}}} = 0$

5. $-\mu M_{un} = -\mu 0 = 0$

Thus, $\dfrac{d M_{nu}}{dt} = 0$ from eq. (1.1-1) and this derivative remains at zero for the remainder of the solution (increasing t).

A similar analysis for eqs. (1.1-2) to (1.1-6) leads to $\dfrac{dM_1}{dt} = \dfrac{dM_2}{dt} = \dfrac{dIL_{10}}{dt} = \dfrac{dT_\alpha}{dt} = \dfrac{dIL_1}{dt} = 0$ so that the six ODE dependent variables remain at the zero ICs throughout the solution.

As a next case, the volumetric monocyte to undifferentiated macrophage rate in the RHS of eq. (1.1-1) is given a positive value $M = 1$ in the parameter list of the main program of Listing 2.1. This nonzero value moves the derivative $\dfrac{dM_{un}}{dt}$ from zero, and therefore $M_{un}(t)$ changes with t. The graphical output is in Figs. 2.2.

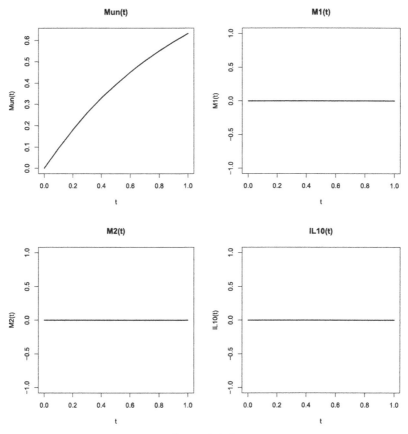

Figure 2.2-1:
Numerical $M_{nu}(t)$, $M_1(t)$, $M_2(t)$, $IL_{10}(t)$ from eqs. (1.1), (1.2), $M = 1$

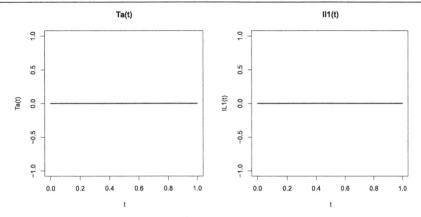

Figure 2.2-2:
Numerical $T_\alpha(t)$, $IL_1(t)$ from eqs. (1.1), (1.2), $M = 1$

Interestingly, $M_1(t)$ to $IL_{(t)}$ remain at the zero ICs even with the variation of $M_{un}(t)$. This result follows from the RHS terms of eqs. (1.1-2) to (1.1-6). That is, eqs. (1.1-2) to (1.1-6) are not coupled to eq. (1.1-1) through $M_{nu}(t)$ if the five ODE dependent variables $M_1(t)$ to $IL_{(t)}$ have zero ICs. Further consideration of this explanation is left as an exercise.

As a third case, the myocyte volumetric rate, λM_c, in eqs. (1.1-5), (1.1-6) is given a nonzero value in the parameter list of the main program of Listing 2.1. Through the coupling between eqs. (1.1-2) to (1.1-6), the derivatives in t are then nonzero and all six ODE dependent variables move away from zero ICs.

For example, if $M = \lambda = M_c = 1$ in the parameter list of Listing 2.1, Figs. 2.3 result.

All six ODE dependent variables move away from zero ICs with $M = \lambda = M_c = 1$.

As a concluding case, IL-1 can have a detrimental effect on cardiac tissue and function. Therefore, medication therapy to reduce the IL-1 concentration is added to eq. (1.1-6) [1].

$$\frac{dIL_1}{dt} = (k_7 M_1 + \lambda M_c)\frac{c}{c + IL_{10}} - d_{IL_1}IL_1 - r_{IL_1}(t) \qquad (2.1\text{-}1)$$

$$r_{IL_1}(t) = r_1 e^{-r2\left(\frac{t - t_0}{t_f - t_0}\right)} \qquad (2.1\text{-}2)$$

The source term to limit $IL_1(t)$, $r_{IL_1}(t)$, in the right hand side (RHS) of eq. (2.1-1) has the exponential form of eq. (2.1-2). It starts (at $t = t_0$) with the value $r_{IL_1}(t = t_0) = r_1 e^0 = r_1$, then decreases in t with $r_2 > 0$ reflecting the reduction in medication (drug) therapy with t for $IL_1(t)$ (from eq. (2.1-1)).

The implementation of eqs. (2.1) follows as extensions of Listing 2.1, 2.2.

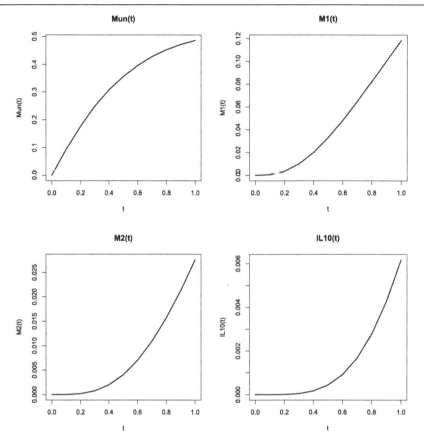

Figure 2.3-1:
Numerical $M_{nu}(t)$, $M_1(t)$, $M_2(t)$, $IL_{10}(t)$ from eqs. (1.1), (1.2), $M = \lambda = M_c = 1$

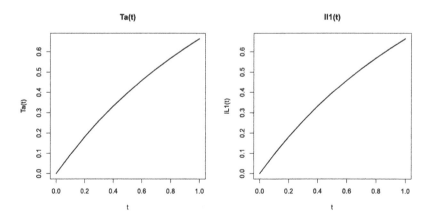

Figure 2.3-2:
Numerical $T_\alpha(t)$, $IL_1(t)$ from eqs. (1.1), (1.2), $M = \lambda = M_c = 1$

2.1.4 Main program with IL-1 control

```
#
# Six ODE post-MI model. Therapeutic
# reduction of IL-1
#
# Delete previous workspaces
  rm(list=ls(all=TRUE))
#
# Access ODE integrator
  library("deSolve");
#
# Access functions for numerical solution
  setwd("g:/myocardial infarction/chap2");
  source("ode1b.R");

     .          .
     .          .
     .          .

  M=1;
 lam=1;
  Mc=1;
  mu=1;
#
# Therapeutic reduction in IL-1
  r1=1;
  r2=1;

     .          .
     .          .
     .          .

#
# ODE integration
  out=lsodes(y=u0,times=tout,func=ode1b,
      sparsetype ="sparseint",rtol=1e-6,
      atol=1e-6,maxord=5);
  nrow(out)
  ncol(out)

     .          .
     .          .
     .          .
```

Listing 2.3: Abbreviated main program for eqs. (1.1), (1.2), with addition of $r_{IL-1}(t)$

We can note

- The ODE routine is ode1b.

  ```
  source("ode1b.R");
  ```

- r_1, r_2 in eq. (2.1-1) are defined numerically.

  ```
  r1=1;
  r2=1;
  ```

- ode1b is called by lsodes

  ```
  out=lsodes(y=u0,times=tout,func=ode1b,
      sparsetype ="sparseint",rtol=1e-6,
      atol=1e-6,maxord=5);
  ```

The abbreviated ODE routine ode1b called by the main program of Listing 2.3 follows.

2.1.5 ODE routine with IL-1 control

```
  ode1b=function(t,u,parms){
#
# Function ode1a computes the t derivative
# vectors of Mun(t),M1(t),M2(t),IL10(t),
# Talpha(t),IL1(t)

      .             .

      .             .

      .             .

    fc=c/(c+IL10);
  rIL1=r1*(1-exp(-r2*(t-t0)/(tf-t0)));

      .             .

      .             .

      .             .
```

```
  IL1t=(k7*M1+lam*Mc)*fc-dIL1*IL1-rIL1;
        .            .
        .            .
        .            .
#
# Return derivative vector
  return(list(c(ut)));
  }
```

Listing 2.4: Abbreviated ODE routine for eqs. (1.1), (1.2), with addition of $r_{IL-1}(t)$

We can note

- The routine is od1b.

  ```
  ode1b=function(t,u,parms){
  ```

- The limiting function of eq. (2.1-2) is coded as rIL1.

  ```
      fc=c/(c+IL10);
    rIL1=r1*(1-exp(-r2*(t-t0)/(tf-t0)));
  ```

- The limiting function is included in eq. (1.1-6).

  ```
  IL1t=(k7*M1+lam*Mc)*fc-dIL1*IL1-rIL1;
  ```

2.1.6 Numerical, graphical output

The output from nrow,ncol called in the main program of Listing 2.3 (dimensions of the solution matrix out from lsodes) follows.

```
[1] 11
```

```
[1] 7
```

As before, the solution matrix out from lsodes is 11×7.

The graphical output of Figs. 2.4.

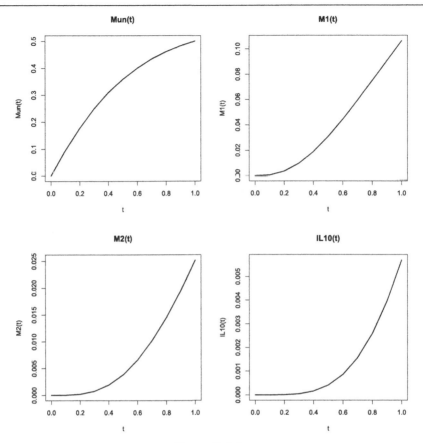

Figure 2.4-1:
Numerical $M_{nu}(t)$, $M_1(t)$, $M_2(t)$, $IL_{10}(t)$ from eqs. (1.1), (1.2), with $r_{IL_1}(t)$

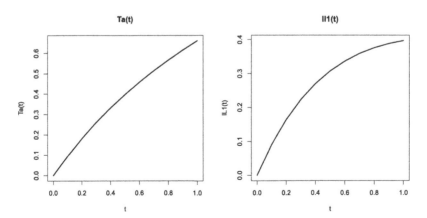

Figure 2.4-2:
Numerical $T_\alpha(t)$, $IL_1(t)$ from eqs. (1.1), (1.2) with $r_{IL_1}(t)$

A comparison of Figs. 2.3-2 and 2.4-2 indicates the effect of r_{IL_1} on IL_1. Variation of r_1 in eq. (2.1-2) will further indicate the contribution of r_{IL_1} to the solution of eq. (1.1-6). This is left as an exercise.

2.2 Summary and conclusions

The six ODE model of eqs. (1.1), (2.1) is implemented in R routines discussed in this chapter that generate numerical solutions $M_{nu}(t)$, $M_1(t)$, $M_2(t)$, $IL_{10}(t)$, $T_\alpha(t)$, $IL_1(t)$ explained in Table 1.1. The use of the IL_1 limiting function $r_{TL_1}(t)$ of eq. (2.1-2) is included to illustrate medicinal therapy to lower the $IL_1(t)$ concentration [1].

The ODE model does not include a spatial aspect. In the following chapters, the spatial variation of the dependent variables of Table 1.1 in the cardiac tissue is included as modeled by partial differential equations (PDEs).

References

[1] C.A. Dinarello, A. Simon, J.W.M. van der Meer, Treating inflammation by blocking interleuken-1 in a broad spectrum of diseases, Nature Reviews Drug Discovery 11 (8) (2012) 633–652.
[2] K. Soetaert, J. Cash, F. Mazzia, Solving Differential Equations in R, Springer-Verlag, Heidelberg, Germany, 2012.

## CHAPTER 3

PDE model formulation and implementation

Introduction

The post-MI ODE model of eqs. (1.1), (1.2) does not include a spatial component for position in the cardiac tissue. In order to include a spatial component, linear diffusion (based on Fick's first, second laws) is added to the ODE model to give a system of six partial differential equations (PDEs), that are stated next.

3.1 Formulation of PDE model

The system of six PDEs follows, resulting from adding diffusion to the RHS of eqs. (1.1) (x is the spatial position in the cardiac tissue).

$$\frac{\partial M_{un}}{\partial t} = D_{M_{un}} \frac{\partial^2 M_{un}}{\partial x^2}$$
$$+ M - k_2 M_{un} \frac{IL_1}{IL_1 + c_{IL_1}} - k_3 M_{un} \frac{T_\alpha}{T_\alpha + c_{T_\alpha}}$$
$$- k_4 M_{un} \frac{IL_{10}}{IL_{10} + c_{IL_{10}}} - \mu M_{un} \tag{3.1-1}$$

$$\frac{\partial M_1}{\partial t} = D_{M_1} \frac{\partial^2 M_1}{\partial x^2}$$
$$+ k_2 M_{un} \frac{IL_1}{IL_1 + c_{IL_1}} + k_3 M_{un} \frac{T_\alpha}{T_\alpha + c_{T_\alpha}}$$
$$+ k_1' M_2 - k_1 M_1 - \mu M_1 \tag{3.1-2}$$

$$\frac{\partial M_2}{\partial t} = D_{M_2} \frac{\partial^2 M_2}{\partial x^2}$$
$$+ k_4 M_{un} \frac{IL_{10}}{IL_{10} + c_{IL_{10}}}$$
$$+ k_1 M_1 - k_1' M_2 - \mu M_2 \tag{3.1-3}$$

$$\frac{\partial IL_{10}}{\partial t} = D_{IL_{10}} \frac{\partial^2 IL_{10}}{\partial x^2}$$
$$+ k_5 M_2 \frac{c_1}{c_1 + IL_{10}} - d_{IL_{10}} IL_{10} \tag{3.1-4}$$

$$\frac{\partial T_\alpha}{\partial t} = D_{T_\alpha} \frac{\partial^2 T_\alpha}{\partial x^2}$$
$$+ (k_6 M_1 + \lambda M_c) \frac{c}{c + IL_{10}} - d_{T_\alpha} T_\alpha \tag{3.1-5}$$

$$\frac{\partial IL_1}{\partial t} = D_{IL_1} \frac{\partial^2 IL_1}{\partial x^2}$$
$$+ (k_7 M_1 + \lambda M_c) \frac{c}{c + IL_{10}} - d_{IL_1} IL_1 \tag{3.1-6}$$

Numerical integration of eqs. (3.1) by the method of lines (MOL)[1] gives the dependent variables

$$M_{un}(x, t), \ M_1(x, t), \ M_2(x, t), \ IL_{10}(x, t), \ T_\alpha(x, t), \ IL_1(x, t)$$

Eqs. (3.1) are second order in x and each requires two boundary conditions (BCs). Homogeneous (no flux) Neumann BCs[2] follow.

$$\frac{\partial M_{nu}(x = x_l, t)}{\partial x} = 0; \quad \frac{\partial M_{nu}(x = x_u, t)}{\partial x} = 0 \tag{3.2-1a,b}$$

$$\frac{\partial M_1(x = x_l, t)}{\partial x} = 0; \quad \frac{\partial M_1(x = x_u, t)}{\partial x} = 0 \tag{3.2-2a,b}$$

$$\frac{\partial M_2(x = x_l, t)}{\partial x} = 0; \quad \frac{\partial M_2(x = x_u, t)}{\partial x} = 0 \tag{3.2-3a,b}$$

$$\frac{\partial IL_{10}(x = x_l, t)}{\partial x} = 0; \quad \frac{\partial IL_{10}(x = x_u, t)}{\partial x} = 0 \tag{3.2-4a,b}$$

$$\frac{\partial T_\alpha(x = x_l, t)}{\partial x} = 0; \quad \frac{\partial T_\alpha(x = x_u, t)}{\partial x} = 0 \tag{3.2-5a,b}$$

$$\frac{\partial IL_1(x = x_l, t)}{\partial x} = 0; \quad \frac{\partial IL_1(x = x_u, t)}{\partial x} = 0 \tag{3.2-6a,b}$$

---

[1]   The method of lines is a general numerical algorithm for PDEs in which the boundary value (spatial) derivatives are replaced with algebraic approximations, in this case, finite differences (FDs). The resulting system of initial value ODEs is then integrated (solved) with a library ODE integrator.

[2]   A Dirichlet BC specifies the dependent variable at the boundary. A Neumann BC specifies the first order spatial derivative of the dependent variable with respect to x at the boundary. A Robin BC includes both the dependent variable and the first order spatial derivative, usually as a linear combination.

Eqs. (3.1) are first order in t and each requires an initial condition (IC) that is a function of x

$$M_{nu}(x, t = 0) = M_{nu,0}(x) \tag{3.3-1}$$

$$M_1(x, t = 0) = M_{1,0}(x) \tag{3.3-2}$$

$$M_2(x, t = 0) = M_{2,0}(x) \tag{3.3-3}$$

$$IL_{10}(x, t = 0) = IL_{10,0}(x) \tag{3.3-4}$$

$$T_\alpha(x, t = 0) = T_{\alpha,0}(x) \tag{3.3-5}$$

$$IL_1(x, t = 0) = IL_{1,0}(x) \tag{3.3-6}$$

Eqs. (3.1), (3.2), (3.3) constitute the six PDE post-MI model that is numerically integrated as follows.

3.2 Implementation of PDE model

The six PDE model is implemented as R routines, starting with a main program.

3.2.1 Main program, test cases

The main program for eqs. (3.1) follows.

```
#
# Six PDE post-MI model
#
# Delete previous workspaces
  rm(list=ls(all=TRUE))
#
# Access ODE integrator
  library("deSolve");
#
# Access functions for numerical solution
  setwd("g:/myocardial infarction/chap3");
  source("pde1a.R");
  source("dss044.R");
#
# Parameters
  nx=41;
```

```
  Mun_0=rep(0,nx);
   M1_0=rep(0,nx);
   M2_0=rep(0,nx);
 IL10_0=rep(0,nx);
   Ta_0=rep(0,nx);
  IL1_0=rep(0,nx);
    k1=1;
   k1p=1;
    k2=1;
    k3=1;
    k4=1;
    k5=1;
    k6=1;
    k7=1;
     c=1;
    c1=1;
  cIL1=1;
   cTa=1;
 cIL10=1;
 dIL10=1;
   dTa=1;
  dIL1=1;
     M=0;
   lam=1;
    Mc=0;
    mu=1;
#
# Diffusivities
   DMun=1;
    DM1=1;
    DM2=1;
  DIL10=1;
    DTa=1;
   DIL1=1;
#
# Independent variable t for PDE integration
  t0=0;tf=1;nout=11;
  tout=seq(from=t0,to=tf,by=(tf-t0)/(nout-1));
#
```

```
# Initial conditions
  xl=0;xu=1;
  x=seq(from=xl,to=xu,by=(xu-xl)/(nx-1));
  u0=rep(0,6*nx);
  for(ix in 1:nx){
    u0[ix]      =Mun_0[ix];
    u0[ix+nx]   = M1_0[ix];
    u0[ix+2*nx] = M2_0[ix];
    u0[ix+3*nx]=IL10_0[ix];
    u0[ix+4*nx]=  Ta_0[ix];
    u0[ix+5*nx]= IL1_0[ix];
  }
  ncall=0;
#
# ODE integration
  out=lsodes(y=u0,times=tout,func=pde1a,
      sparsetype ="sparseint",rtol=1e-6,
      atol=1e-6,maxord=5);
  nrow(out)
  ncol(out)
#
# Arrays for plotting numerical solution
   Munp=matrix(0,nrow=nx,ncol=nout);
    M1p=matrix(0,nrow=nx,ncol=nout);
    M2p=matrix(0,nrow=nx,ncol=nout);
  IL10p=matrix(0,nrow=nx,ncol=nout);
    Tap=matrix(0,nrow=nx,ncol=nout);
   IL1p=matrix(0,nrow=nx,ncol=nout);
  for(it in 1:nout){
    for(ix in 1:nx){
       Munp[ix,it]=out[it,ix+1];
        M1p[ix,it]=out[it,ix+1+nx];
        M2p[ix,it]=out[it,ix+1+2*nx];
      IL10p[ix,it]=out[it,ix+1+3*nx];
        Tap[ix,it]=out[it,ix+1+4*nx];
       IL1p[ix,it]=out[it,ix+1+5*nx];
    }
  }
#
```

```
# Plot PDE solutions
# 2D
#
# Mun
# par(mfrow=c(1,1));
  par(mfrow=c(2,2));
  matplot(x=x,y=Munp,type="l",xlab="x",ylab="Mun(x,t)",
          xlim=c(t0,tf),lty=1,main="Mun(x,t)",lwd=2,
          col="black");
#
# M1
  matplot(x=x,y=M1p,type="l",xlab="x",ylab="M1(x,t)",
          xlim=c(t0,tf),lty=1,main="M1(x,t)",lwd=2,
          col="black");
#
# M2
  matplot(x=x,y=M2p,type="l",xlab="x",ylab="M2(x,t)",
          xlim=c(t0,tf),lty=1,main="M2(x,t)",lwd=2,
          col="black");
#
# IL10
  matplot(x=x,y=IL10p,type="l",xlab="x",ylab="IL10(x,t)",
          xlim=c(t0,tf),lty=1,main="IL10(x,t)",lwd=2,
          col="black");
#
# Ta
  matplot(x=x,y=Tap,type="l",xlab="x",ylab="Ta(x,t)",
          xlim=c(t0,tf),lty=1,main="Ta(x,t)",lwd=2,
          col="black");
#
# IL1
  matplot(x=x,y=IL1p,type="l",xlab="x",ylab="IL1(x,t)",
          xlim=c(t0,tf),lty=1,main="IL1(x,t)",lwd=2,
          col="black");
#
# 3D
  par(mfrow=c(2,2));
  if(Mc==0){zul=0.1};
  if(Mc> 0){zul=1.1};
```

```
persp(x,tout,Munp,theta=60,phi=45,
     xlim=c(xl,xu),ylim=c(t0,tf),zlim=c(0,zul),
     xlab="x",ylab="t",zlab="Mun(x,t)");
persp(x,tout,M1p,theta=60,phi=45,
     xlim=c(xl,xu),ylim=c(t0,tf),zlim=c(0,zul),
     xlab="x",ylab="t",zlab="M1(x,t)");
persp(x,tout,M2p,theta=60,phi=45,
     xlim=c(xl,xu),ylim=c(t0,tf),zlim=c(0,zul),
     xlab="x",ylab="t",zlab="M2(x,t)");
persp(x,tout,IL10p,theta=60,phi=45,
     xlim=c(xl,xu),ylim=c(t0,tf),zlim=c(0,zul),
     xlab="x",ylab="t",zlab="IL10(x,t)");
persp(x,tout,Tap,theta=60,phi=45,
     xlim=c(xl,xu),ylim=c(t0,tf),zlim=c(0,zul),
     xlab="x",ylab="t",zlab="Ta(x,t)");
persp(x,tout,IL1p,theta=60,phi=45,
     xlim=c(xl,xu),ylim=c(t0,tf),zlim=c(0,zul),
     xlab="x",ylab="t",zlab="IL1(x,t)");
```

Listing 3.1: Main program for eqs. (3.1), (3.2), (3.3)

We can note the following details about Listing 3.1 (with some repetition of the discussion of Listing 2.1 so that the following discussion is self contained).

- Previous workspaces are deleted.

```
#
# Six PDE post-MI model
#
# Delete previous workspaces
  rm(list=ls(all=TRUE))
```

- The R ODE integrator library deSolve is accessed [1]. Then the directory with the files for the solution of eqs. (3.1), (3.2), (3.3) is designated. Note that setwd (set working directory) uses / rather than the usual \.

```
#
# Access ODE integrator
  library("deSolve");
#
# Access functions for numerical solution
```

```
setwd("g:/myocardial infarction/chap3");
source("pde1a.R");
source("dss044.R");
```

pde1a is the routine for eqs. (3.1), (3.2), (3.3) discussed subsequently. dss044 is a library routine for second order spatial derivatives (listed in book Appendix A).

- The model parameters are specified numerically.

```
#
# Parameters
  nx=41;
  Mun_0=rep(0,nx);
   M1_0=rep(0,nx);
   M2_0=rep(0,nx);
 IL10_0=rep(0,nx);
   Ta_0=rep(0,nx);
  IL1_0=rep(0,nx);
   k1=1;
  k1p=1;
   k2=1;
   k3=1;
   k4=1;
   k5=1;
   k6=1;
   k7=1;
    c=1;
   c1=1;
 cIL1=1;
  cTa=1;
 cIL10=1;
 dIL10=1;
   dTa=1;
  dIL1=1;
    M=0;
  lam=1;
   Mc=0;
   mu=1;
```

- The diffusivities in eqs. (3.1) are defined numerically.

```
#
# Diffusivities
   DMun=1;
    DM1=1;
    DM2=1;
  DIL10=1;
    DTa=1;
   DIL1=1;
```

- An interval in *t* is defined for 11 output points, so that `tout=0,0.1,...,1`.

```
#
# Independent variable t for PDE integration
  t0=0;tf=1;nout=11;
  tout=seq(from=t0,to=tf,by=(tf-t0)/(nout-1));
```

- ICs (3.3) are programmed.

```
#
# Initial conditions
  xl=0;xu=1;
  x=seq(from=xl,to=xu,by=(xu-xl)/(nx-1));
  u0=rep(0,6*nx);
  for(ix in 1:nx){
    u0[ix]      =Mun_0[ix];
    u0[ix+nx]   = M1_0[ix];
    u0[ix+2*nx] = M2_0[ix];
    u0[ix+3*nx]=IL10_0[ix];
    u0[ix+4*nx]=  Ta_0[ix];
    u0[ix+5*nx]= IL1_0[ix];
  }
  ncall=0
```

Also, the counter for the calls to pde1a is initialized.
- The system of $6 * 41 = 246$ ODEs is integrated by the library integrator lsodes (available in deSolve, [1]). As expected, the inputs to lsodes are the ODE/MOL function, pde1a, the IC vector u0, and the vector of output values of *t*, tout. The length of u0 (246) informs lsodes how many ODEs are to be integrated. func,y,times are reserved names.

```
#
# ODE integration
  out=lsodes(y=u0,times=tout,func=pde1a,
      sparsetype ="sparseint",rtol=1e-6,
      atol=1e-6,maxord=5);
  nrow(out)
  ncol(out)
```

nrow,ncol confirm the dimensions of out.

* $M_{nu}(x,t)$, $M_1(x,t)$, $M_2(x,t)$, $IL_{10}(x,t)$, $T_\alpha(x,t)$, $IL_1(x,t)$ are placed in arrays for numerical and graphical output.

```
#
# Arrays for plotting numerical solution
   Mun=matrix(0,nrow=nx,ncol=nout);
    M1=matrix(0,nrow=nx,ncol=nout);
    M2=matrix(0,nrow=nx,ncol=nout);
  IL10=matrix(0,nrow=nx,ncol=nout);
    Ta=matrix(0,nrow=nx,ncol=nout);
   IL1=matrix(0,nrow=nx,ncol=nout);
  for(it in 1:nout){
    for(ix in 1:nx){
       Mun[ix,it]=out[it,ix+1];
        M1[ix,it]=out[it,ix+1+nx];
        M2[ix,it]=out[it,ix+1+2*nx];
      IL10[ix,it]=out[it,ix+1+3*nx];
        Ta[ix,it]=out[it,ix+1+4*nx];
       IL1[ix,it]=out[it,ix+1+5*nx];
    }
  }
```

The offset +1 is required because the first element of the solution vectors in out is the value of t and the 2 to 247 elements are the $(6)(41) - 246$ values of $M_{nu}(x,t)$, $M_1(x,t)$, $M_2(x,t)$, $IL_{10}(x,t)$, $T_\alpha(x,t)$, $IL_1(x,t)$. These dimensions from the preceding calls to nrow,ncol are confirmed in the subsequent output.

* $M_{nu}(x,t)$, $M_1(x,t)$, $M_2(x,t)$, $IL_{10}(x,t)$, $T_\alpha(x,t)$, $IL_1(x,t)$ are plotted in 2D with the matplot utility. par(mfrow=c(2,2)) specifies a 2×2 matrix of plots on a page of graphical output.

```
#
# Plot PDE solutions
# 2D
#
# Mun
# par(mfrow=c(1,1));
  par(mfrow=c(2,2));
  matplot(x=x,y=Mun,type="l",xlab="x",ylab="Mun(x,t)",
          xlim=c(t0,tf),lty=1,main="Mun(x,t)",lwd=2,
          col="black");
#
# M1
  matplot(x=x,y=M1,type="l",xlab="x",ylab="M1(x,t)",
          xlim=c(t0,tf),lty=1,main="M1(x,t)",lwd=2,
          col="black");
#
# M2
  matplot(x=x,y=M2,type="l",xlab="x",ylab="M2(x,t)",
          xlim=c(t0,tf),lty=1,main="M2(x,t)",lwd=2,
          col="black");
#
# IL10
  matplot(x=x,y=IL10,type="l",xlab="x",ylab="IL10(x,t)",
          xlim=c(t0,tf),lty=1,main="IL10(x,t)",lwd=2,
          col="black");
#
# Ta
  matplot(x=x,y=Ta,type="l",xlab="x",ylab="Ta(x,t)",
          xlim=c(t0,tf),lty=1,main="Ta(x,t)",lwd=2,
          col="black");
#
# IL1
  matplot(x=x,y=IL1,type="l",xlab="x",ylab="IL1(x,t)",
          xlim=c(t0,tf),lty=1,main="IL1(x,t)",lwd=2,
          col="black");
```

For $M_c = 0$ there is no cross coupling of the PDEs as explained in Chapter 2. The five PDE dependent variables $M_1(x,t)$, $M_2(x,t)$, $IL_{10}(x,t)$, $T_\alpha(x,t)$, $IL_1(x,t)$ remain at the ICs.

For $M_c > 0$ cross coupling of the PDEs is demonstrated in the numerical and graphical output.

- $M_{nu}(x,t)$, $M_1(x,t)$, $M_2(x,t)$, $IL_{10}(x,t)$, $T_\alpha(x,t)$, $IL_1(x,t)$ are plotted in 3D with the persp utility.

```
#
# 3D
  par(mfrow=c(2,2));
  zul=1.1;
  persp(x,tout,Mun,theta=60,phi=45,
        xlim=c(xl,xu),ylim=c(t0,tf),zlim=c(0,zul),
        xlab="x",ylab="t",zlab="Mun(x,t)");
  persp(x,tout,M1,theta=60,phi=45,
        xlim=c(xl,xu),ylim=c(t0,tf),zlim=c(0,zul),
        xlab="x",ylab="t",zlab="M1(x,t)");
  persp(x,tout,M2,theta=60,phi=45,
        xlim=c(xl,xu),ylim=c(t0,tf),zlim=c(0,zul),
        xlab="x",ylab="t",zlab="M2(x,t)");
  persp(x,tout,IL10,theta=60,phi=45,
        xlim=c(xl,xu),ylim=c(t0,tf),zlim=c(0,zul),
        xlab="x",ylab="t",zlab="IL10(x,t)");
  persp(x,tout,Ta,theta=60,phi=45,
        xlim=c(xl,xu),ylim=c(t0,tf),zlim=c(0,zul),
        xlab="x",ylab="t",zlab="Ta(x,t)");
  persp(x,tout,IL1,theta=60,phi=45,
        xlim=c(xl,xu),ylim=c(t0,tf),zlim=c(0,zul),
        xlab="x",ylab="t",zlab="IL1(x,t)");
```

The scaling of the z axis, zlim=c(0,zul), is defined with zul=1.1; since the automatic scaling does not work correctly (persp gives an error message)) when there is no variation in the vertical variable ($u_{un}(x,t)$ remains at the IC with $M = 0$).

This completes the discussion of the main program for eqs. (3.1), (3.2), (3.3). The ODE/MOL routine pde1a called by lsodes from the main program (Listing 3.1) for the numerical integration of eqs. (3.1), (3.2), (3.3) is next.

3.2.2 ODE/MOL routine

pde1a called in the main program of Listing 3.1 follows.

```
  pde1a=function(t,u,parms){
#
# Function ode1a computes the t derivative
# vectors of Mun(x,t),M1(x,t),M2(x,t),IL10(x,t),
# Ta(x,t),IL1(x,t)
#
# One vector to six vectors
   Mun=rep(0,nx);M1=rep(0,nx); M2=rep(0,nx);
  IL10=rep(0,nx);Ta=rep(0,nx);IL1=rep(0,nx);
  for(ix in 1:nx){
     Mun[ix]=u[ix];
      M1[ix]=u[ix+nx];
      M2[ix]=u[ix+2*nx];
    IL10[ix]=u[ix+3*nx];
      Ta[ix]=u[ix+4*nx];
     IL1[ix]=u[ix+5*nx];
  }
#
# Munx,M1x,M2x,IL10x,Tax,IL1x
   Munx=rep(0,nx);M1x=rep(0,nx); M2x=rep(0,nx);
  IL10x=rep(0,nx);Tax=rep(0,nx);IL1x=rep(0,nx);
#
# BCs
   Munx[1]=0;  Munx[nx]=0;
    M1x[1]=0;   M1x[nx]=0;
    M2x[1]=0;   M2x[nx]=0;
  IL10x[1]=0;IL10x[nx]=0;
    Tax[1]=0;   Tax[nx]=0;
   IL1x[1]=0;  IL1x[nx]=0;
#
# Munxx,M1xx,M2xx,IL10xx,Taxx,IL1xx
  nl=2;nu=2;
  Munxx=dss044(xl,xu,nx, Mun, Munx,nl,nu);
   M1xx=dss044(xl,xu,nx,   M1,   M1x,nl,nu);
   M2xx=dss044(xl,xu,nx,   M2,   M2x,nl,nu);
 IL10xx=dss044(xl,xu,nx,IL10,IL10x,nl,nu);
   Taxx=dss044(xl,xu,nx,   Ta,  Tax,nl,nu);
  IL1xx=dss044(xl,xu,nx,  IL1,  IL1x,nl,nu);
#
```

```
# MOL/ODEs
   Munt=rep(0,nx); M1t=rep(0,nx); M2t=rep(0,nx);
  IL10t=rep(0,nx); Tat=rep(0,nx);IL1t=rep(0,nx);
  for(ix in 1:nx){
   fIL1=k2*Mun[ix]*IL1[ix]/(IL1[ix]+cIL1);
    fTa=k3*Mun[ix]*Ta[ix]/(Ta[ix]+cTa);
  fIL10=k4*Mun[ix]*IL10[ix]/(IL10[ix]+cIL10);
    fc1=k5*M2[ix]*c1/(c1+IL10[ix]);
      fc=c/(c+IL10[ix]);
  Munt[ix]=DMun*Munxx[ix]+M-fIL1-fTa-fIL10-mu*Mun[ix];
    M1t[ix]=DM1*M1xx[ix]+fIL1+fTa+k1p*M2[ix]-k1*M1[ix]-mu*M1[ix];
    M2t[ix]=DM2*M2xx[ix]+fIL10+k1*M1[ix]-k1p*M2[ix]-mu*M2[ix];
  IL10t[ix]=DIL10*IL10xx[ix]+fc1-dIL10*IL10[ix];
    Tat[ix]=DTa*Taxx[ix]+(k6*M1[ix]+lam*Mc)*fc-dTa*Ta[ix];
   IL1t[ix]=DIL1*IL1xx[ix]+(k7*M1[ix]+lam*Mc)*fc-dIL1*IL1[ix];
  }
#
# Six vectors to one vector
  ut=rep(0,6*nx);
  for(ix in 1:nx){
    ut[ix]      = Munt[ix];
    ut[ix+nx]   =  M1t[ix];
    ut[ix+2*nx]=   M2t[ix];
    ut[ix+3*nx]=IL10t[ix];
    ut[ix+4*nx] = Tat[ix];
    ut[ix+5*nx]= IL1t[ix];
  }
#
# Increment calls to pde1a
  ncall <<- ncall+1;
#
# Return derivative vector
  return(list(c(ut)));
  }
```

Listing 3.2: ODE/MOL routine pde1a for eqs. (3.1), (3.2), (3.3)

We can note the following details about pde1a (with some repetition of the discussion of Listing 2.2 so that the following discussion is self contained)

- The function is defined.

```
#
# Six PDE post-MI model
#
# Delete previous workspaces
  rm(list=ls(all=TRUE))
```

t is the current value of *t* in eqs. (3.1). u is the $(6)(41) = 246$-vector of ODE dependent variables. parm is an argument to pass parameters to pde1a (unused, but required in the argument list). The arguments must be listed in the order stated to properly interface with lsodes called in the main program of Listing 3.1. The derivative vector of the LHS of eqs. (3.1) is calculated and returned to lsodes as explained subsequently.

- Vector u is placed in six vectors to facilitate the programming of eqs. (3.1).

```
#
# One vector to six vectors
  Mun=rep(0,nx);M1=rep(0,nx); M2=rep(0,nx);
  IL10=rep(0,nx);Ta=rep(0,nx);IL1=rep(0,nx);
  for(ix in 1:nx){
    Mun[ix]=u[ix];
     M1[ix]=u[ix+nx];
     M2[ix]=u[ix+2*nx];
   IL10[ix]=u[ix+3*nx];
     Ta[ix]=u[ix+4*nx];
    IL1[ix]=u[ix+5*nx];
  }
```

- BCs (3.2) are programmed (homogeneous Neumann BCs).

```
#
# Munx,M1x,M2x,IL10x,Tax,IL1x
  Munx=rep(0,nx);M1x=rep(0,nx); M2x=rep(0,nx);
  IL10x=rep(0,nx);Tax=rep(0,nx);IL1x=rep(0,nx);
#
# BCs
  Munx[1]=0; Munx[nx]=0;
   M1x[1]=0;  M1x[nx]=0;
   M2x[1]=0;  M2x[nx]=0;
```

```
   IL10x[1]=0;IL10x[nx]=0;
     Tax[1]=0;   Tax[nx]=0;
   IL1x[1]=0;  IL1x[nx]=0;
```

• The second order derivatives in eqs. (3.1) are computed with dss044.

$$\frac{\partial^2 M_{un}(x,t)}{\partial x^2}, \frac{\partial^2 M_1(x,t)}{\partial x^2}, \frac{\partial^2 M_2(x,t)}{\partial x^2}, \frac{\partial^2 IL_{10}(x,t)}{\partial x^2}, \frac{\partial^2 T_\alpha(x,t)}{\partial x^2}, \frac{\partial^2 IL_1(x,t)}{\partial x^2}$$

```
#
# Munxx,M1xx,M2xx,IL10xx,Taxx,IL1xx
  nl=2;nu=2;
  Munxx=dss044(xl,xu,nx, Mun, Munx,nl,nu);
   M1xx=dss044(xl,xu,nx,  M1,  M1x,nl,nu);
   M2xx=dss044(xl,xu,nx,  M2,  M2x,nl,nu);
 IL10xx=dss044(xl,xu,nx,IL10,IL10x,nl,nu);
   Taxx=dss044(xl,xu,nx,  Ta,  Tax,nl,nu);
  IL1xx=dss044(xl,xu,nx, IL1, IL1x,nl,nu);
```

nl=2, nu=2 specify Neumann BCs at $x = x_l, x = x_u$, respectively.

• Eqs. (3.1) are programmed, starting with the nonlinear rate functions.

```
#
# MOL/ODEs
  Munt=rep(0,nx); M1t=rep(0,nx); M2t=rep(0,nx);
  IL10t=rep(0,nx); Tat=rep(0,nx);IL1t=rep(0,nx);
  for(ix in 1:nx){
   fIL1=k2*Mun[ix]*IL1[ix]/(IL1[ix]+cIL1);
    fTa=k3*Mun[ix]*Ta[ix]/(Ta[ix]+cTa);
  fIL10=k4*Mun[ix]*IL10[ix]/(IL10[ix]+cIL10);
    fc1=k5*M2[ix]*c1/(c1+IL10[ix]);
     fc=c/(c+IL10[ix]);
```

The programming of the nonlinear functions is explained after Listing 2.2. These examples demonstrate the straightforward use of nonlinear functions numerically, which would be difficult to accomplish analytically. Also, since these functions are computed in the programming of the ODE/MOLs, they can be displayed graphically (plotted) as well as the PDE dependent variables, which gives a direct indication of the contribution of the nonlinear functions.

- Eqs. (3.1) are programmed in the ODE/MOL format (within the `for` in `ix`).

```
 Munt[ix]=DMun*Munxx[ix]+M-fIL1-fTa-fIL10-mu*Mun[ix];
  M1t[ix]=DM1*M1xx[ix]+fIL1+fTa+k1p*M2[ix]-k1*M1[ix]-mu*M1[ix];
  M2t[ix]=DM2*M2xx[ix]+fIL10+k1*M1[ix]-k1p*M2[ix]-mu*M2[ix];
IL10t[ix]=DIL10*IL10xx[ix]+fc1-dIL10*IL10[ix];
  Tat[ix]=DTa*Taxx[ix]+(k6*M1[ix]+lam*Mc)*fc-dTa*Ta[ix];
 IL1t[ix]=DIL1*IL1xx[ix]+(k7*M1[ix]+lam*Mc)*fc-dIL1*IL1[ix];
}
```

For example, the left hand side (LHS) derivative $\dfrac{\partial M_{un}}{\partial t}$ in eq. (3.1-1) is programmed as `Munt[ix]`, and the RHS

$$D_{M_{un}}\frac{\partial^2 M_{un}}{\partial x^2}$$

$$+M - k_2 M_{un}\frac{IL_1}{IL_1 + c_{IL_1}} - k_3 M_{un}\frac{T_\alpha}{T_\alpha + c_{T_\alpha}}$$

$$-k_4 M_{un}\frac{IL_{10}}{IL_{10} + c_{IL_{10}}} - \mu M_{un}$$

is programmed as `DMun*Munxx[ix]+M-fIL1-fTa-fIL10-mu*Mun[ix];`
- The $(6)(41) = 246$ ODE derivatives are placed in the vector `ut` for return to `lsodes` to take the next step in t along the solution.

```
#
# Six vectors to one vector
  ut=rep(0,6*nx);
  for(ix in 1:nx){
    ut[ix]      = Munt[ix];
    ut[ix+nx]   =  M1t[ix];
    ut[ix+2*nx]=   M2t[ix];
    ut[ix+3*nx]=IL10t[ix];
    ut[ix+4*nx] = Tat[ix];
    ut[ix+5*nx]= IL1t[ix];
}
```

- The counter for the calls to `pde1a` is incremented and returned to the main program of Listing 3.1 by `<<-`.

```
#
# Increment calls to pde1a
  ncall <<- ncall+1;
```

- The vector ut is returned as a list as required by lsodes. c is the R vector utility. The final } concludes pde1a.

```
#
# Return derivative vector
  return(list(c(ut)));
  }
```

This completes the discussion of pde1a. The output from the main program of Listing 3.1 and ODE/MOL routine pde1a of Listing 3.2 is considered next.

3.2.3 Numerical, graphical output

The output from nrow,ncol called in the main program of Listing 3.1 (dimensions of the solution matrix out from lsodes) follows.

```
[1] 11
```

```
[1] 247
```

The row dimension of the solution matrix out from lsodes is 11 corresponding to the definition of the *t* values of tout.

```
#
# Independent variable for ODE integration
  t0=0;tf=1;nout=11;
  tout=seq(from=t0,to=tf,by=(tf-t0)/(nout-1));
```

The column dimension of out is 247 corresponding to the 246 ODE dependent variables $M_{un}(x, t)$ to $IL_1(x, t)$ of eqs. (3.1) and the independent variable *t* for the $(6)(41) = 246$ dependent variables.

As a **first case**, the graphical output of Figs. 3.1 indicates that the six dependent variables remain at the homogeneous ICs of eqs. (3.3).

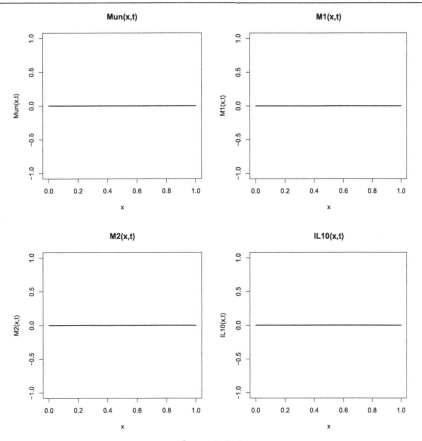

Figure 3.1-1:
Numerical $M_{nu}(x,t)$, $M_1(x,t)$, $M_2(x,t)$, $IL_{10}(x,t)$ from eqs. (3.1), (3.2), (3.3), $M = M_c = 0$, 2D

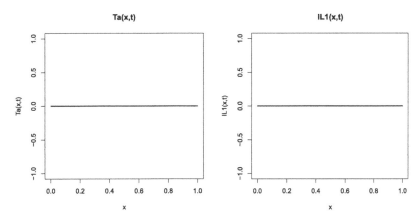

Figure 3.1-2:
Numerical $T_{\alpha}(x,t)$, $IL_1(x,t)$ from eqs. (3.1), (3.2), (3.3), $M = M_c = 0$, 2D

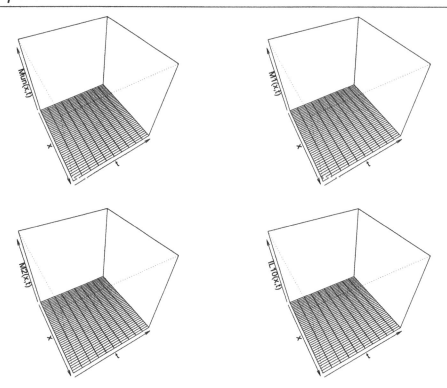

Figure 3.1-3:
Numerical $M_{nu}(x,t)$, $M_1(x,t)$, $M_2(x,t)$, $IL_{10}(x,t)$ from eqs. (3.1), (3.2), (3.3), $M = M_c = 0$, 3D

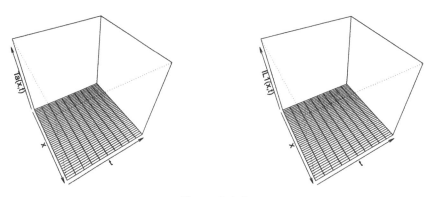

Figure 3.1-4:
Numerical $T_\alpha(x,t)$, $IL_1(x,t)$ from eqs. (3.1), (3.2), (3.3), $M = M_c = 0$, 3D

The six PDE dependent variables remain at the homogeneous (zero) ICs since all of the RHS terms of the PDEs are zero. This is an important special case since a departure from the zero ICs would indicate a programming error.

As an example of a PDE dependent variable that remains at the IC, $M_{un}(x, t)$ from eq. (3.1-1) is analyzed in Table 3.1 for zero ICs.

Table 3.1: Explanation of RHS terms in eq. (3.1-1) for zero ICs

1. $D_{un} \dfrac{\partial^2 u_{un}(x, t)}{\partial^2 x} \approx D_{un} \left(\dfrac{u_{un}(x + \Delta x, t) - 2u_{un}(x, t) + u_{un}(x - \Delta x, t)}{\Delta x^2} \right) = D_{un}(0 - 2(0) + 0)/\Delta x^2 = 0.$
   At the boundaries $x = x_l$, $x = x_u$, this diffusion term is zero through application of BCs (3.2-1a,b).

2. $M = 0$ set as a parameter in the main program of Listing 3.1.
   For $x = x_l$, $x_l + \Delta x$, ..., $x = x_u$ (stepping through the interval in x, $x_l \le x \le x_u$ with increment Δx),

3. $-k_2 M_{un}(x, t = 0) \dfrac{IL_1(x, t = 0)}{IL_1(x, t = 0) + c_{IL_1}} = -k_2 0 \dfrac{0}{0 + c_{IL_1}} = 0$

4. $-k_3 M_{un}(x, t = 0) \dfrac{T_\alpha(x, t = 0)}{T_\alpha(x, t = 0) + c_{T_\alpha}} = -k_3 0 \dfrac{0}{0 + c_{T_\alpha}} = 0$

5. $-k_4 M_{un}(x, t = 0) \dfrac{IL_{10}(x, t = 0)}{IL_{10}(x, t = 0) + c_{IL_{10}}} = -k_4 0 \dfrac{0}{0 + c_{IL_{10}}} = 0$

6. $-\mu M_{un}(x, t = 0) = -\mu 0 = 0$

Thus, $\dfrac{\partial M_{nu}(x, t = 0)}{\partial t} = 0$ from eq. (3.1-1) and this derivative remains at zero for the remainder of the solution (increasing t).

A similar analysis for eqs. (3.1-2) to (3.1-6) leads to $\dfrac{\partial M_1(x, t = 0)}{\partial t} = \dfrac{\partial M_2(x, t = 0)}{\partial t} = \dfrac{\partial IL_{10}(x, t = 0)}{\partial t} = \dfrac{\partial T_\alpha(x, t = 0)}{\partial t} = \dfrac{\partial IL_1(x, t = 0)}{\partial t} = 0$ so that the six PDE dependent variables in t remain at the zero ICs throughout the solution.

As a **second case**, the volumetric monocyte to undifferentiated macrophage rate in the RHS of eq. (3.1-1) is given a positive value $M = 1$ in the parameter list of the main program of Listing 3.1. This nonzero value moves the derivative $\dfrac{\partial M_{un}(x, t)}{\partial t}$ from zero, and therefore $M_{un}(x, t)$ changes with t. The graphical output is in Figs. 3.2.

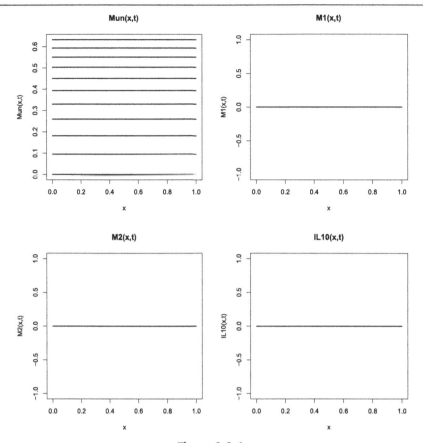

Figure 3.2-1:
Numerical $M_{nu}(x,t)$, $M_1(x,t)$, $M_2(x,t)$, $IL_{10}(x,t)$ from eqs. (3.1), (3.2), (3.3), $M = 1$, $M_c = 0$, 2D

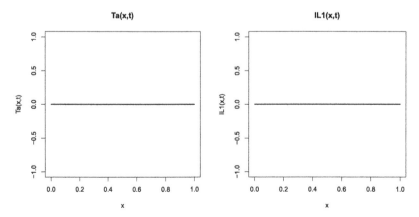

Figure 3.2-2:
Numerical $T_\alpha(x,t)$, $IL_1(x,t)$ from eqs. (3.1), (3.2), (3.3), $M = 1$, $M_c = 0$, 2D

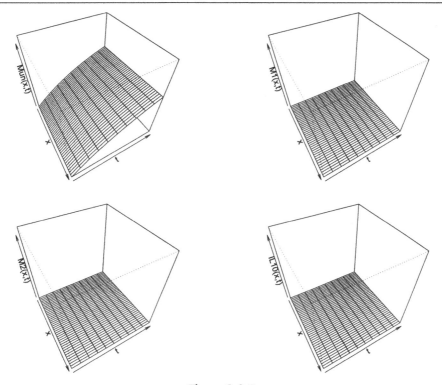

Figure 3.2-3:
Numerical $M_{nu}(x, t)$, $M_1(x, t)$, $M_2(x, t)$, $IL_{10}(x, t)$ from eqs. (3.1), (3.2), (3.3), $M = 1$, $M_c = 0$, 3D

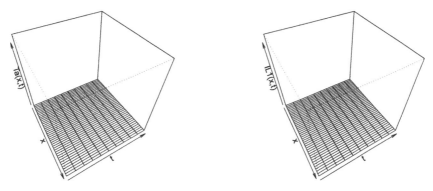

Figure 3.2-4:
Numerical $T_\alpha(x, t)$, $IL_1(x, t)$ from eqs. (3.1), (3.2), (3.3), $M = 1$, $M_c = 0$, 3D

$M_{nu}(x, t)$ is uniform in x, which follows from the homogeneous IC (3.3-1) and zero Neumann BCs (3.2-1a,b). This is an important test since any variation of $M_{nu}(x, t)$ with x would indicate a programming error.

Interestingly, $M_1(x, t)$ to $IL(x, t)$ remain at the zero ICs even with the variation of $M_{un}(x, t)$. This result follows from the RHS terms of eqs. (3.1-2) to (3.1-6). That is, eqs. (3.1-2) to (3.1-6) are not coupled to eq. (3.1-1) through $M_{nu}(x, t)$ if the five PDE dependent variables $M_1(x, t)$ to $IL(x, t)$ have zero ICs. Further consideration of this explanation (PDE decoupling) is left as an exercise.

As a **third case**, the myocyte volumetric rate, λM_c, in eqs. (3.1-5), (3.1-6) is given a nonzero value in the parameter list of the main program of Listing 3.1 with $M = 0, \lambda = M_c = 1$. The derivatives $\dfrac{\partial T_\alpha(x, t)}{\partial t}$, $\dfrac{\partial IL_1(x, t)}{\partial t}$ are nonzero and $T_\alpha(x, t)$, $IL_1(x, t)$ move away from zero ICs.

The graphical output is in Figs. 3.3.

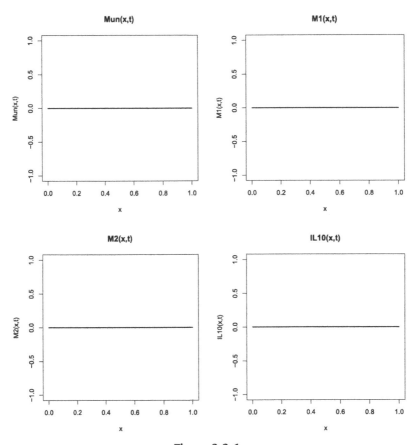

Figure 3.3-1:
Numerical $M_{nu}(x, t)$, $M_1(x, t)$, $M_2(x, t)$, $IL_{10}(x, t)$ from eqs. (3.1), (3.2), (3.3), $M = 0, M_c = 1$, 2D

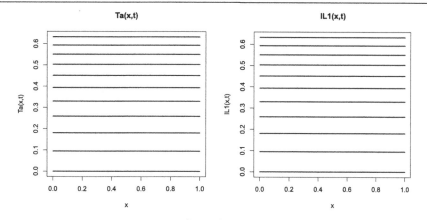

Figure 3.3-2:
Numerical $T_\alpha(x,t)$, $IL_1(x,t)$ from eqs. (3.1), (3.2), (3.3), $M = 0$, $M_c = 1$, 2D

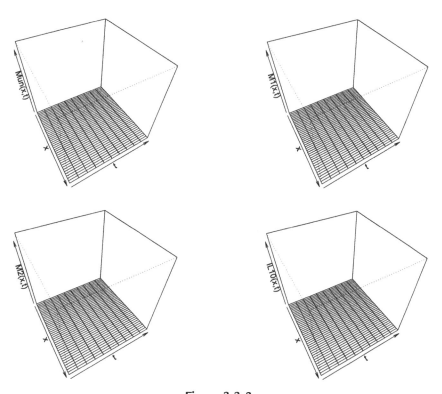

Figure 3.3-3:
Numerical $M_{nu}(x,t)$, $M_1(x,t)$, $M_2(x,t)$, $IL_{10}(x,t)$ from eqs. (3.1), (3.2), (3.3), $M = 0$, $M_c = 1$, 3D

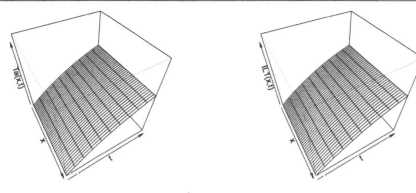

Figure 3.3-4:
Numerical $T_\alpha(x, t)$, $IL_1(x, t)$ from eqs. (3.1), (3.2), (3.3), $M = 0$, $M_c = 1$, 3D

$T_\alpha(x, t)$, $IL_1(x, t)$ move away from zero ICs with $M = 0, \lambda = M_c = 1$.

As a **fourth case**, the monocyte rate M in eq. (3.1-1) and the myocyte rate λM_c in eqs. (1.1-5), (1.1-6) are given nonzero values in the parameter list of the main program of Listing 3.1 with $M = 1, \lambda = M_c = 1$. The derivatives $\dfrac{\partial M_{un}(x, t)}{\partial t}, \dfrac{\partial M_1(x, t)}{\partial t}, \dfrac{\partial M_2(x, t)}{\partial t}, \dfrac{\partial IL_{10}(x, t)}{\partial t},$ $\dfrac{\partial T_\alpha(x, t)}{\partial t}, \dfrac{\partial IL_1(x, t)}{\partial t}$ are nonzero and $M_{nu}(x, t), M_1(x, t), M_2(x, t), IL_{10}(x, t), T_\alpha(x, t),$ $IL_1(x, t)$ move away from zero ICs.

The graphical output is in Figs. 3.4.

$M_{nu}(x, t), M_1(x, t)\ M_2(x, t), IL_{10}(x, t)\ T_\alpha(x, t), IL_1(x, t)$ move away from zero ICs with $M = 1, \lambda = M_c = 1$.

The scaling of the vertical axis in the calls to persp is uniform from zul=1.1, zlim=c(0,zul) in

```
#
# 3D
  par(mfrow=c(2,2));
  zul=1.1;
  persp(x,tout,Mun,theta=60,phi=45,
       xlim=c(xl,xu),ylim=c(t0,tf),zlim=c(0,zul),
       xlab="x",ylab="t",zlab="Mun(x,t)");
            .              .
            .              .
            .              .
```

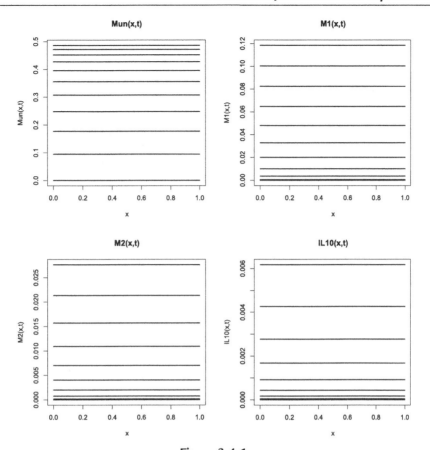

Figure 3.4-1:
Numerical $M_{nu}(x,t)$, $M_1(x,t)$, $M_2(x,t)$, $IL_{10}(x,t)$ from eqs. (3.1), (3.2), (3.3), $M = M_c = 1$, 2D

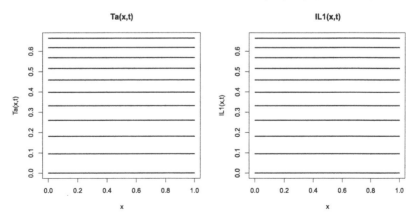

Figure 3.4-2:
Numerical $T_\alpha(x,t)$, $IL_1(x,t)$ from eqs. (3.1), (3.2), (3.3), $M = M_c = 1$, 2D

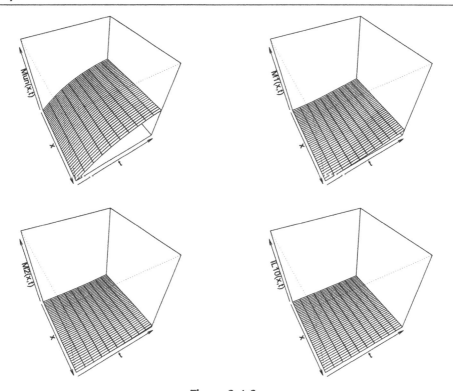

Figure 3.4-3:
Numerical $M_{nu}(x,t)$, $M_1(x,t)$, $M_2(x,t)$, $IL_{10}(x,t)$ from eqs. (3.1), (3.2), (3.3), $M = M_c = 1$, 3D

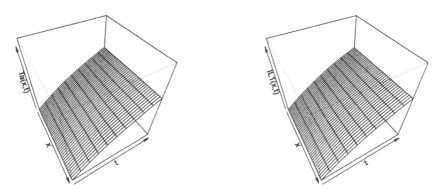

Figure 3.4-4:
Numerical $T_\alpha(x,t)$, $IL_1(x,t)$ from eqs. (3.1), (3.2), (3.3), $M = M_c = 1$, 3D

```
persp(x,tout,IL1,theta=60,phi=45,
      xlim=c(xl,xu),ylim=c(t0,tf),zlim=c(0,zul),
      xlab="x",ylab="t",zlab="IL1(x,t)");
```

Thus, the dependent variables with small vertical variations, e.g., $M_1(x, 0)$, $M_2(x, t)$, $IL_{10}(x, t)$ appear to be near constant at zero ICs. The variation in 3D can be demonstrated by removing `zlim=c(0,zul)` in the calls to `persp`. This is left as an exercise.

As a **fifth case**, the monocyte rate M in eq. (3.1-1) and the myocyte rate λM_c in eqs. (1.1-5), (1.1-6) are given a spatial variation, $M(x, t)$, $M_c(x, t)$. That is, the post-MI varies spatially in the cardiac tissue.

The changes in the main program of Listing 3.1 are indicated in Listing 3.3.

```
#
# Six PDE post-MI model
#
# Delete previous workspaces
  rm(list=ls(all=TRUE))
#
# Access ODE integrator
  library("deSolve");
#
# Access functions for numerical solution
  setwd("g:/myocardial infarction/chap3");
  source("pde1b.R");
  source("dss044.R");
#
# Parameters
  nx=41;
  Mun_0=rep(0,nx);
   M1_0=rep(0,nx);
   M2_0=rep(0,nx);
 IL10_0=rep(0,nx);
   Ta_0=rep(0,nx);
  IL1_0=rep(0,nx);
   k1=1;
  k1p=1;
   k2=1;
```

```
   k3=1;
   k4=1;
   k5=1;
   k6=1;
   k7=1;
    c=1;
   c1=1;
 cIL1=1;
  cTa=1;
cIL10=1;
dIL10=1;
  dTa=1;
 dIL1=1;
  lam=1;
   mu=1;
#
# Select case
  ncase=2;
#
# Spatially variable monocytes, myocytes
   M=rep(0,nx);
  Mc=rep(0,nx);
#
# ncase=1
  if(ncase==1){
    for(ix in 1:nx){
       M[ix]=1;
     Mc[ix]=1;
    }
 }
#
# ncase=2
  if(ncase==2){
    nx2=(nx-1)/2+1;
    cat(sprintf("\n nx = %3d nx/2 = %3d\n",nx,nx2));
    cat(sprintf("\n  ix    M(x)   Mc(x)\n"));
    for(ix in 1:nx){
      if(ix==nx2){
        M[ix]=1;
```

```
          Mc[ix]=1;
        }
        if(ix==(nx2-1)){
          M[ix]=0.5;
         Mc[ix]=0.5;
        }
        if(ix==(nx2+1)){
          M[ix]=0.5;
         Mc[ix]=0.5;
        }
        cat(sprintf("%4d%8.2f%8.2f\n",ix,M[ix],Mc[ix]));
    }
  }
#
# Diffusivities
   DMun=1;
    DM1=1;
    DM2=1;
  DIL10=1;
    DTa=1;
   DIL1=1;
#
# Independent variable t for PDE integration
  t0=0;tf=1;nout=11;
  tout=seq(from=t0,to=tf,by=(tf-t0)/(nout-1));
#
# Initial conditions
  xl=0;xu=1;
  x=seq(from=xl,to=xu,by=(xu-xl)/(nx-1));
  u0=rep(0,6*nx);
  for(ix in 1:nx){
    u0[ix]      =Mun_0[ix];
    u0[ix+nx]   = M1_0[ix];
    u0[ix+2*nx] = M2_0[ix];
    u0[ix+3*nx]=IL10_0[ix];
    u0[ix+4*nx]=  Ta_0[ix];
    u0[ix+5*nx]= IL1_0[ix];
  }
```

```
  ncall=0;
#
# ODE integration
  out=lsodes(y=u0,times=tout,func=pde1b,
      sparsetype ="sparseint",rtol=1e-6,
      atol=1e-6,maxord=5);
  nrow(out)
  ncol(out)
#
# Arrays for plotting numerical solution
   Mun=matrix(0,nrow=nx,ncol=nout);
    M1=matrix(0,nrow=nx,ncol=nout);
    M2=matrix(0,nrow=nx,ncol=nout);
  IL10=matrix(0,nrow=nx,ncol=nout);
    Ta=matrix(0,nrow=nx,ncol=nout);
   IL1=matrix(0,nrow=nx,ncol=nout);
  for(it in 1:nout){
    for(ix in 1:nx){
       Mun[ix,it]=out[it,ix+1];
        M1[ix,it]=out[it,ix+1+nx];
        M2[ix,it]=out[it,ix+1+2*nx];
      IL10[ix,it]=out[it,ix+1+3*nx];
        Ta[ix,it]=out[it,ix+1+4*nx];
       IL1[ix,it]=out[it,ix+1+5*nx];
    }
  }
#
# Plot PDE solutions
# 2D
#
# Mun
# par(mfrow=c(1,1));
  par(mfrow=c(2,2));
  matplot(x=x,y=Mun,type="l",xlab="x",ylab="Mun(x,t)",
          xlim=c(t0,tf),lty=1,main="Mun(x,t)",lwd=2,
          col="black");
#
# M1
```

```
  matplot(x=x,y=M1,type="l",xlab="x",ylab="M1(x,t)",
          xlim=c(t0,tf),lty=1,main="M1(x,t)",lwd=2,
          col="black");
#
# M2
  matplot(x=x,y=M2,type="l",xlab="x",ylab="M2(x,t)",
          xlim=c(t0,tf),lty=1,main="M2(x,t)",lwd=2,
          col="black");
#
# IL10
  matplot(x=x,y=IL10,type="l",xlab="x",ylab="IL10(x,t)",
          xlim=c(t0,tf),lty=1,main="IL10(x,t)",lwd=2,
          col="black");
#
# Ta
  matplot(x=x,y=Ta,type="l",xlab="x",ylab="Ta(x,t)",
          xlim=c(t0,tf),lty=1,main="Ta(x,t)",lwd=2,
          col="black");
#
# IL1
  matplot(x=x,y=IL1,type="l",xlab="x",ylab="IL1(x,t)",
          xlim=c(t0,tf),lty=1,main="IL1(x,t)",lwd=2,
          col="black");
#
# 3D
  par(mfrow=c(2,2));
  persp(x,tout,Mun,theta=60,phi=45,
        xlim=c(xl,xu),ylim=c(t0,tf),
        xlab="x",ylab="t",zlab="Mun(x,t)");
  persp(x,tout,M1,theta=60,phi=45,
        xlim=c(xl,xu),ylim=c(t0,tf),
        xlab="x",ylab="t",zlab="M1(x,t)");
  persp(x,tout,M2,theta=60,phi=45,
        xlim=c(xl,xu),ylim=c(t0,tf),
        xlab="x",ylab="t",zlab="M2(x,t)");
  persp(x,tout,IL10,theta=60,phi=45,
        xlim=c(xl,xu),ylim=c(t0,tf),
        xlab="x",ylab="t",zlab="IL10(x,t)");
```

```
persp(x,tout,Ta,theta=60,phi=45,
      xlim=c(xl,xu),ylim=c(t0,tf),
      xlab="x",ylab="t",zlab="Ta(x,t)");
persp(x,tout,IL1,theta=60,phi=45,
      xlim=c(xl,xu),ylim=c(t0,tf),
      xlab="x",ylab="t",zlab="IL1(x,t)");
```

Listing 3.3: Main program for eqs. (3.1), (3.2), (3.3) with $M(x,t)$, $M_c(x,t)$

We can note the following details about Listing 3.3 (with some repetition of the discussion of Listings 2.1 and 3.1) so that the following discussion is self contained).

- Previous workspaces are deleted.

```
#
# Six PDE post-MI model
#
# Delete previous workspaces
  rm(list=ls(all=TRUE))
```

- The R ODE integrator library deSolve is accessed [1]. Then the directory with the files for the solution of eqs. (3.1), (3.2), (3.3) is designated. Note that setwd (set working directory) uses / rather than the usual \.

```
#
# Access ODE integrator
  library("deSolve");
#
# Access functions for numerical solution
  setwd("g:/myocardial infarction/chap3");
  source("pde1b.R");
  source("dss044.R");
```

pde1b is the routine for eqs. (3.1), (3.2), (3.3) discussed subsequently. dss044 is a library routine for second order spatial derivatives (listed in book Appendix A).
- The model parameters are specified numerically.

```
#
# Parameters
  nx=41;
  Mun_0=rep(0,nx);
```

```
      M1_0=rep(0,nx);
      M2_0=rep(0,nx);
    IL10_0=rep(0,nx);
      Ta_0=rep(0,nx);
     IL1_0=rep(0,nx);
       k1=1;
      k1p=1;
       k2=1;
       k3=1;
       k4=1;
       k5=1;
       k6=1;
       k7=1;
        c=1;
       c1=1;
     cIL1=1;
      cTa=1;
    cIL10=1;
    dIL10=1;
      dTa=1;
     dIL1=1;
      lam=1;
       mu=1;
```

- Two cases are programmed for $M(x,t)$, $M_c(x,t)$.

```
#
# Select case
  ncase=1;
#
# Spatially variable monocytes, myocytes
   M=rep(0,nx);
  Mc=rep(0,nx);
#
# ncase=1
  if(ncase==1){
    for(ix in 1:nx){
      M[ix]=1;
     Mc[ix]=1;
```

```
    }
#
# ncase=2
  if(ncase==2){
    nx2=(nx-1)/2+1;
    cat(sprintf("\n nx = %3d nx/2 = %3d\n",nx,nx2));
    cat(sprintf("\n  ix    M(x)    Mc(x)\n"));
    for(ix in 1:nx){
      if(ix==nx2){
        M[ix]=1;
       Mc[ix]=1;
      }
      if(ix==(nx2-1)){
        M[ix]=0.5;
       Mc[ix]=0.5;
      }
      if(ix==(nx2+1)){
        M[ix]=0.5;
       Mc[ix]=0.5;
      }
      cat(sprintf("%4d%8.2f%8.2f\n",ix,M[ix],Mc[ix]));
    }
  }
```

This programming requires some additional explanation.

- After the selection of ncase, $M(x,t)$, $M_c(x,t)$ are defined on a spatial grid in x of nx points.

```
#
# Select case
  ncase=1;
#
# Spatially variable monocytes, myocytes
    M=rep(0,nx);
   Mc=rep(0,nx);
```

– For ncase=1, $M(x,t) = M_c(x,t) = 1$ corresponding to the preceding fourth case. Therefore, the six PDE solutions are uniform in x as confirmed in Figs. 3.5-1,2,3,4.

```
#
# ncase=1
  if(ncase==1){
    for(ix in 1:nx){
      M[ix]=1;
      Mc[ix]=1;
    }
```

Also, the correspondence of the solutions in Figs. 3.4-1,2,3,4 ($M = M_c = 1$) and Figs. 3.5-1,2,3,4 ($M(x,t) = M_c(x,t) = 1$) is clear.

– For ncase=2, the monocyte and myocyte generation rates, $M(x,t)$ in eq. (3.1-1) and $M_c(x,t)$ in eqs. (3.1-5,6), are functions of x corresponding to a spatial variation of the infarction in the cardiac tissue.

```
#
# ncase=2
  if(ncase==2){
    nx2=(nx-1)/2+1;
    cat(sprintf("\n nx = %3d nx/2 = %3d\n",nx,nx2));
    cat(sprintf("\n  ix    M(x)    Mc(x)\n"));
    for(ix in 1:nx){
      if(ix==nx2){
        M[ix]=1;
        Mc[ix]=1;
      }
      if(ix==(nx2-1)){
        M[ix]=0.5;
        Mc[ix]=0.5;
      }
      if(ix==(nx2+1)){
        M[ix]=0.5;
        Mc[ix]=0.5;
      }
      cat(sprintf("%4d%8.2f%8.2f\n",ix,M[ix],Mc[ix]));
    }
  }
```

As an example, the spatial variation of $M(x, t)$ and $M_c(x, t)$ is a triangular pulse in x centered at the midpoint of the interval $x_l \leq x \leq x_u$, that is, $x_2 = (x_u - x_l)/2$, with the index ix = nx2=(nx-1)/2+1. The specific values of the spatial variation functions are (with $\Delta x = (x_u - x_l)/(n_x - 1)$)

$$x_l \leq x \leq x_2 - 2\Delta x;\ M(x, t) = M_c(x, t) = 0$$
$$x = x_2 - \Delta x;\ M(x, t) = M_c(x, t) = 0.5$$
$$x = x_2;\ M(x, t) = M_c(x, t) = 1$$
$$x = x_2 + \Delta x;\ M(x, t) = M_c(x, t) = 0.5$$
$$x_2 + 2\Delta x \leq x \leq x_u;\ M(x, t) = M_c(x, t) = 0$$

- The remainder of the main program is the same as in Listing 3.1 with the exception that the ODE/MOL routine is pde1b called by lsodes, and the vertical scaling of the 3D plots (from persp, zul=1.1) is not used.

```
    .
    .
    .
#
# ODE integration
  out=lsodes(y=u0,times=tout,func=pde1b,
      sparsetype ="sparseint",rtol=1e-6,
      atol=1e-6,maxord=5);
  nrow(out)
  ncol(out)
    .
    .
    .
#
# 3D
  par(mfrow=c(2,2));
  persp(x,tout,Mun,theta=60,phi=45,
        xlim=c(xl,xu),ylim=c(t0,tf),
        xlab="x",ylab="t",zlab="Mun(x,t)");
    .
    .
    .
  persp(x,tout,IL1,theta=60,phi=45,
```

```
              xlim=c(xl,xu),ylim=c(t0,tf),
              xlab="x",ylab="t",zlab="IL1(x,t)");
```

The ODE/PDE routine follows.

```
  pde1b=function(t,u,parms){
#
# Function pde1b computes the t derivative
# vectors of Mun(x,t),M1(x,t),M2(x,t),IL10(x,t),
# Ta(x,t),IL1(x,t)
#
# One vector to six vectors
  Mun=rep(0,nx);M1=rep(0,nx); M2=rep(0,nx);
  IL10=rep(0,nx);Ta=rep(0,nx);IL1=rep(0,nx);
  for(ix in 1:nx){
     Mun[ix]=u[ix];
      M1[ix]=u[ix+nx];
      M2[ix]=u[ix+2*nx];
    IL10[ix]=u[ix+3*nx];
      Ta[ix]=u[ix+4*nx];
     IL1[ix]=u[ix+5*nx];
  }
#
# Munx,M1x,M2x,IL10x,Tax,IL1x
  Munx=rep(0,nx);M1x=rep(0,nx); M2x=rep(0,nx);
  IL10x=rep(0,nx);Tax=rep(0,nx);IL1x=rep(0,nx);
#
# BCs
  Munx[1]=0; Munx[nx]=0;
   M1x[1]=0;  M1x[nx]=0;
   M2x[1]=0;  M2x[nx]=0;
  IL10x[1]=0;IL10x[nx]=0;
   Tax[1]=0;  Tax[nx]=0;
   IL1x[1]=0; IL1x[nx]=0;
#
# Munxx,M1xx,M2xx,IL10xx,Taxx,IL1xx
  nl=2;nu=2;
```

```
  Munxx=dss044(xl,xu,nx, Mun, Munx,nl,nu);
   M1xx=dss044(xl,xu,nx,  M1,  M1x,nl,nu);
   M2xx=dss044(xl,xu,nx,  M2,  M2x,nl,nu);
 IL10xx=dss044(xl,xu,nx,IL10,IL10x,nl,nu);
   Taxx=dss044(xl,xu,nx,  Ta,  Tax,nl,nu);
  IL1xx=dss044(xl,xu,nx, IL1, IL1x,nl,nu);
#
# MOL/ODEs
   Munt=rep(0,nx); M1t=rep(0,nx); M2t=rep(0,nx);
  IL10t=rep(0,nx); Tat=rep(0,nx);IL1t=rep(0,nx);
  for(ix in 1:nx){
   fIL1=k2*Mun[ix]*IL1[ix]/(IL1[ix]+cIL1);
    fTa=k3*Mun[ix]*Ta[ix]/(Ta[ix]+cTa);
  fIL10=k4*Mun[ix]*IL10[ix]/(IL10[ix]+cIL10);
    fc1=k5*M2[ix]*c1/(c1+IL10[ix]);
     fc=c/(c+IL10[ix]);
   Munt[ix]=DMun*Munxx[ix]+M[ix]-fIL1-fTa-fIL10-mu*Mun[ix];
    M1t[ix]=DM1*M1xx[ix]+fIL1+fTa+k1p*M2[ix]-k1*M1[ix]-mu*M1[ix];
    M2t[ix]=DM2*M2xx[ix]+fIL10+k1*M1[ix]-k1p*M2[ix]-mu*M2[ix];
  IL10t[ix]=DIL10*IL10xx[ix]+fc1-dIL10*IL10[ix];
    Tat[ix]=DTa*Taxx[ix]+(k6*M1[ix]+lam*Mc[ix])*fc-dTa*Ta[ix];
   IL1t[ix]=DIL1*IL1xx[ix]+(k7*M1[ix]+lam*Mc[ix])*fc-dIL1*IL1[ix];
   }
#
# Six vectors to one vector
  ut=rep(0,6*nx);
  for(ix in 1:nx){
    ut[ix]      = Munt[ix];
    ut[ix+nx]   =  M1t[ix];
    ut[ix+2*nx]=   M2t[ix];
    ut[ix+3*nx]=IL10t[ix];
    ut[ix+4*nx] = Tat[ix];
    ut[ix+5*nx]= IL1t[ix];
   }
#
# Increment calls to pde1b
  ncall <<- ncall+1;
```

```
#
# Return derivative vector
  return(list(c(ut)));
  }
```

Listing 3.4: ODE/MOL routine pde1b for eqs. (3.1), (3.2), (3.3) with $M(x,t)$, $M_c(x,t)$

pde1b, is the same as pde1a of Listing 3.2 except that the monocyte and myocyte rates are changed from M, M_c to $M(x,t)$, $M_c(x,t)$.

Listing 3.2

```
  Munt[ix]=DMun*Munxx[ix]+M-fIL1-fTa-fIL10-mu*Mun[ix];
                .                            .
                .                            .
                .                            .

  Tat[ix]=DTa*Taxx[ix]+(k6*M1[ix]+lam*Mc)*fc-dTa*Ta[ix];
  IL1t[ix]=DIL1*IL1xx[ix]+(k7*M1[ix]+lam*Mc)*fc-dIL1*IL1[ix];
```

Listing 3.4

```
  Munt[ix]=DMun*Munxx[ix]+M[ix]-fIL1-fTa-fIL10-mu*Mun[ix];
                .                            .
                .                            .
                .                            .

  Tat[ix]=DTa*Taxx[ix]+(k6*M1[ix]+lam*Mc[ix])*fc-dTa*Ta[ix];
  IL1t[ix]=DIL1*IL1xx[ix]+(k7*M1[ix]+lam*Mc[ix])*fc-dIL1*IL1[ix];
```

The numerical output for ncase=1 is the same as for the solution matrix out returned by lsodes for the solution constant in x.

```
[1]  11
```

```
[1]  247
```

The graphical output is in Figs. 3.5-1,2,3,4 for ncase=1.

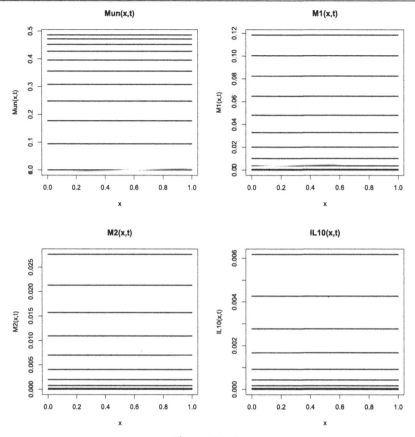

Figure 3.5-1:
Numerical $M_{nu}(x,t)$, $M_1(x,t)$, $M_2(x,t)$, $IL_{10}(x,t)$ from eqs. (3.1), (3.2), (3.3), $M(x,t)$, $M_c(x,t)$, ncase=1, 2D

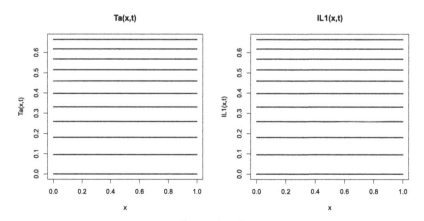

Figure 3.5-2:
Numerical $T_\alpha(x,t)$, $IL_1(x,t)$ from eqs. (3.1), (3.2), (3.3), $M(x,t)$, $M_c(x,t)$, ncase=1, 2D

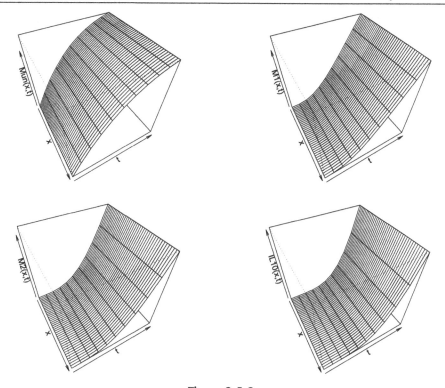

Figure 3.5-3:
Numerical $M_{nu}(x,t)$, $M_1(x,t)$, $M_2(x,t)$, $IL_{10}(x,t)$ from eqs. (3.1), (3.2), (3.3), $M(x,t)$, $M_c(x,t)$, ncase=1, 3D

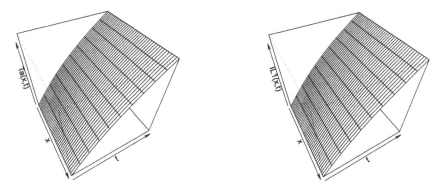

Figure 3.5-4:
Numerical $T_\alpha(x,t)$, $IL_1(x,t)$ from eqs. (3.1), (3.2), (3.3), $M(x,t)$, $M_c(x,t)$, ncase=1, 3D

$M_{nu}(x,t)$, $M_1(x,t)$ $M_2(x,t)$, $IL_{10}(x,t)$ $T_\alpha(x,t)$, $IL_1(x,t)$ move away from zero ICs with $M = 1, \lambda = M_c = 1$.

The numerical output for ncase=2 follows.

```
nx =  41 nx/2 =  21

 ix    M(x)    Mc(x)
  1    0.00    0.00
  2    0.00    0.00
  3    0.00    0.00
          .       .
          .       .
          .       .
 17    0.00    0.00
 18    0.00    0.00
 19    0.00    0.00
 20    0.50    0.50
 21    1.00    1.00
 22    0.50    0.50
 23    0.00    0.00
 24    0.00    0.00
 25    0.00    0.00
          .       .
          .       .
          .       .
 39    0.00    0.00
 40    0.00    0.00
 41    0.00    0.00
```

[1] 11

[1] 247

The variable $M(x,t)$, $M_c(x,t)$ clearly conform to the triangular pulse programmed in Listing 3.3.

The graphical output is in Figs. 3.5-5,6,7,8 for ncase=2.

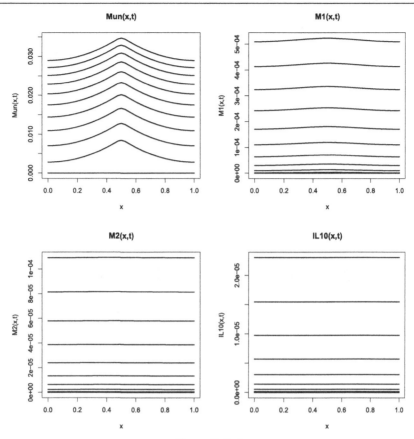

Figure 3.5-5:
Numerical $M_{nu}(x,t)$, $M_1(x,t)$, $M_2(x,t)$, $IL_{10}(x,t)$ from eqs. (3.1), (3.2), (3.3), $M(x,t)$, $M_c(x,t)$, ncase=2, 2D

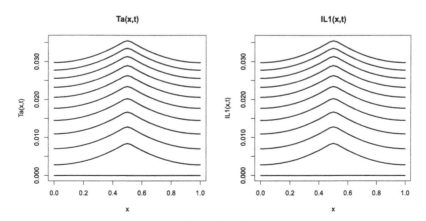

Figure 3.5-6:
Numerical $T_\alpha(x,t)$, $IL_1(x,t)$ from eqs. (3.1), (3.2), (3.3), $M(x,t)$, $M_c(x,t)$, ncase=2, 2D

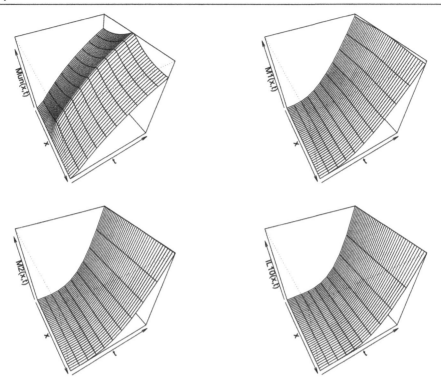

Figure 3.5-7:
Numerical $M_{nu}(x,t)$, $M_1(x,t)$, $M_2(x,t)$, $IL_{10}(x,t)$ from eqs. (3.1), (3.2), (3.3), $M(x,t)$, $M_c(x,t)$, ncase=2, 3D

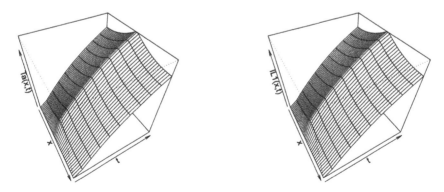

Figure 3.5-8:
Numerical $T_\alpha(x,t)$, $IL_1(x,t)$ from eqs. (3.1), (3.2), (3.3), $M(x,t)$, $M_c(x,t)$, ncase=2, 3D

$M_{nu}(x,t)$, $M_1(x,t)$ $M_2(x,t)$, $IL_{10}(x,t)$ $T_\alpha(x,t)$, $IL_1(x,t)$ move away from zero ICs with $M(x,t)$, $M_c(x,t)$ (programmed in Listing 3.1).

The effect of the triangular $M(x,t)$, $M_c(x,t)$ programmed for `ncase=2` in Listing 3.3 is clear in Figs. 3.5-5,6,7,8. The solutions do not reach a steady state since $M(x,t)$, $M_c(x,t)$ continue to contribute to the t derivatives of eqs. (3.1-1), (3.1-5,6).

3.3 Summary and conclusions

The ODE model of eqs. (1.1), (1.2) is extended in this chapter to include a spatial aspect expressed as a system of PDEs, eqs. (3.1), (3.2), (3.3). The monocyte generation rate in eq. (3.1-1) and myocyte generation rate in eqs. (3.1-5,6) are extended from constants to functions of space so that PDE solutions reflect a post-MI spatial response of the six PDE dependent variables. Experimentation with the two generation rates gives an indication of the effects of spatial post-MI.

The features of the ODE/PDE solutions, e.g., in Figs. 2.4, 3.5, are determined by the left hand side (LHS) derivatives in t of eqs. (1.1), (3.1). These derivatives are computed and displayed in the next chapter for additional insight into the ODE/PDE models.

References

[1] K. Soetaert, J. Cash, F. Mazzia, Solving Differential Equations in R, Springer-Verlag, Heidelberg, Germany, 2012.

CHAPTER 4

PDE model temporal derivative analysis

Introduction

The features of the PDE solutions of eqs. (3.1), (3.2), (3.3), e.g., in Figs. 3.5, are determined by the left hand side (LHS) derivatives in t of eqs. (3.1). These derivatives are computed and displayed in this chapter for additional insight into the PDE modeling of post-MI.

4.1 LHS time derivative analysis of the PDE model

The calculation and display of the PDE LHS time derivative vector

$$\frac{\partial M_{un}(x,t)}{\partial t}, \ \frac{\partial M_1(x,t)}{\partial t}, \ \frac{\partial M_2(x,t)}{\partial t}, \ \frac{\partial IL_{10}(x,t)}{\partial t}, \ \frac{\partial T_\alpha(x,t)}{\partial t}, \ \frac{\partial IL_1(x,t)}{\partial t}$$

of eqs. (3.1) is implemented through a main program and ODE/MOL routine [1]. The main program is in Listing 3.3 with the addition of the following code for the calculation of the PDE t derivative vector.

4.1.1 Addition to the main program

The following coding for the calculation and display of the PDE t derivative vector is added to the end of the main program of Listing 3.3.

```
#
# Six PDE post-MI model
# RHS t derivatives
#
# Delete previous workspaces
  rm(list=ls(all=TRUE))
#
# Access ODE integrator
  library("deSolve");
#
# Access functions for numerical solution
```

```
setwd("g:/myocardial infarction/chap4");
source("pde1a.R");
source("dss004.R");
source("dss044.R");
      .         .
      .         .
      .         .

#
# Munx,M1x,M2x,IL10x,Tax,IL1x
  Munx=matrix(0,nrow=nx,ncol=nout);
   M1x=matrix(0,nrow=nx,ncol=nout);
   M2x=matrix(0,nrow=nx,ncol=nout);
 IL10x=matrix(0,nrow=nx,ncol=nout);
   Tax=matrix(0,nrow=nx,ncol=nout);
  IL1x=matrix(0,nrow=nx,ncol=nout);
  for(it in 1:nout){
   Munx[,it]=dss004(xl,xu,nx, Mun[,it]);
    M1x[,it]=dss004(xl,xu,nx,  M1[,it]);
    M2x[,it]=dss004(xl,xu,nx,  M2[,it]);
  IL10x[,it]=dss004(xl,xu,nx,IL10[,it]);
    Tax[,it]=dss004(xl,xu,nx,  Ta[,it]);
   IL1x[,it]=dss004(xl,xu,nx, IL1[,it]);
  }
#
# BCs
  for(it in 1:nout){
    Munx[1,it]=0; Munx[nx,it]=0;
     M1x[1,it]=0;  M1x[nx,it]=0;
     M2x[1,it]=0;  M2x[nx,it]=0;
   IL10x[1,it]=0;IL10x[nx,it]=0;
     Tax[1,it]=0;  Tax[nx,it]=0;
    IL1x[1,it]=0; IL1x[nx,it]=0;
  }
#
# Munxx,M1xx,M2xx,IL10xx,Taxx,IL1xx
  Munxx=matrix(0,nrow=nx,ncol=nout);
   M1xx=matrix(0,nrow=nx,ncol=nout);
   M2xx=matrix(0,nrow=nx,ncol=nout);
 IL10xx=matrix(0,nrow=nx,ncol=nout);
```

```
     Taxx=matrix(0,nrow=nx,ncol=nout);
   IL1xx=matrix(0,nrow=nx,ncol=nout);
   nl=2;nu=2;
   for(it in 1:nout){
      Munxx[,it]=dss044(xl,xu,nx, Mun[,it], Munx[,it],nl,nu);
       M1xx[,it]=dss044(xl,xu,nx,  M1[,it],  M1x[,it],nl,nu);
       M2xx[,it]=dss044(xl,xu,nx,  M2[,it],  M2x[,it],nl,nu);
     IL10xx[,it]=dss044(xl,xu,nx,IL10[,it],IL10x[,it],nl,nu);
       Taxx[,it]=dss044(xl,xu,nx,  Ta[,it],  Tax[,it],nl,nu);
      IL1xx[,it]=dss044(xl,xu,nx, IL1[,it],  IL1x[,it],nl,nu);
   }
#
# PDE t derivatives
   Munt=matrix(0,nrow=nx,ncol=nout);
    M1t=matrix(0,nrow=nx,ncol=nout);
    M2t=matrix(0,nrow=nx,ncol=nout);
  IL10t=matrix(0,nrow=nx,ncol=nout);
    Tat=matrix(0,nrow=nx,ncol=nout);
   IL1t=matrix(0,nrow=nx,ncol=nout);
   for(it in 1:nout){
   for(ix in 1:nx){
    fIL1=k2*Mun[ix,it]*IL1[ix,it]/(IL1[ix,it]+cIL1);
     fTa=k3*Mun[ix,it]*Ta[ix,it]/(Ta[ix,it]+cTa);
   fIL10=k4*Mun[ix,it]*IL10[ix,it]/(IL10[ix,it]+cIL10);
     fc1=k5*M2[ix,it]*c1/(c1+IL10[ix,it]);
      fc=c/(c+IL10[ix,it]);
   Munt[ix,it]=DMun*Munxx[ix,it]+M[ix]-fIL1-fTa-fIL10-mu*Mun[ix,it];
     M1t[ix,it]=DM1*M1xx[ix,it]+fIL1+fTa+k1p*M2[ix,it]-k1*M1[ix,it]-
                mu*M1[ix,it];
     M2t[ix,it]=DM2*M2xx[ix,it]+fIL10+k1*M1[ix,it]-k1p*M2[ix,it]-
                mu*M2[ix,it];
   IL10t[ix,it]=DIL10*IL10xx[ix,it]+fc1-dIL10*IL10[ix,it];
     Tat[ix,it]=DTa*Taxx[ix,it]+(k6*M1[ix,it]+lam*Mc[ix])*fc-
                dTa*Ta[ix,it];
    IL1t[ix,it]=DIL1*IL1xx[ix,it]+(k7*M1[ix,it]+lam*Mc[ix])*fc-
                dIL1*IL1[ix,it];
   }
   }
#
```

```
# Plot PDE t derivatives
# 2D
#
# Mun
# par(mfrow=c(1,1));
  par(mfrow=c(2,2));
  matplot(x=x,y=Munt,type="l",xlab="x",ylab="Munt(x,t)",
          xlim=c(t0,tf),lty=1,main="Munt(x,t)",lwd=2,
          col="black");
#
# M1
  matplot(x=x,y=M1t,type="l",xlab="x",ylab="M1t(x,t)",
          xlim=c(t0,tf),lty=1,main="M1t(x,t)",lwd=2,
          col="black");
#
# M2
  matplot(x=x,y=M2t,type="l",xlab="x",ylab="M2t(x,t)",
          xlim=c(t0,tf),lty=1,main="M2t(x,t)",lwd=2,
          col="black");
#
# IL10
  matplot(x=x,y=IL10t,type="l",xlab="x",ylab="IL10t(x,t)",
          xlim=c(t0,tf),lty=1,main="IL10t(x,t)",lwd=2,
          col="black");
#
# Ta
  matplot(x=x,y=Tat,type="l",xlab="x",ylab="Tat(x,t)",
          xlim=c(t0,tf),lty=1,main="Tat(x,t)",lwd=2,
          col="black");
#
# IL1
  matplot(x=x,y=IL1t,type="l",xlab="x",ylab="IL1t(x,t)",
          xlim=c(t0,tf),lty=1,main="IL1t(x,t)",lwd=2,
          col="black");
#
# 3D
  par(mfrow=c(2,2));
  persp(x,tout,Munt,theta=60,phi=45,
        xlim=c(xl,xu),ylim=c(t0,tf),
```

```
          xlab="x",ylab="t",zlab="Munt(x,t)");
  persp(x,tout,M1t,theta=60,phi=45,
        xlim=c(xl,xu),ylim=c(t0,tf),
        xlab="x",ylab="t",zlab="M1t(x,t)");
  persp(x,tout,M2t,theta=60,phi=45,
        xlim=c(xl,xu),ylim=c(t0,tf),
        xlab="x",ylab="t",zlab="M2t(x,t)");
  persp(x,tout,IL10t,theta=60,phi=45,
        xlim=c(xl,xu),ylim=c(t0,tf),
        xlab="x",ylab="t",zlab="IL10t(x,t)");
  persp(x,tout,Tat,theta=60,phi=45,
        xlim=c(xl,xu),ylim=c(t0,tf),
        xlab="x",ylab="t",zlab="Tat(x,t)");
  persp(x,tout,IL1t,theta=60,phi=45,
        xlim=c(xl,xu),ylim=c(t0,tf),
        xlab="x",ylab="t",zlab="IL1t(x,t)");
```

Listing 4.1: Addition to the main program for the calculation and display of the PDE LHS *t* derivative vectors

The code in Listing 4.1 follows from Listing 3.4, with the spatial variation indexed by `ix` extended to spatial and temporal variations indexed by `ix` and `it`. We can note the following details about Listing 4.1.

- The ODE/MOL routine is `pde1a`, discussed subsequently.

```
#
# Six PDE post-MI model
# RHS t derivatives
#
# Delete previous workspaces
  rm(list=ls(all=TRUE))
#
# Access ODE integrator
  library("deSolve");
#
# Access functions for numerical solution
  setwd("g:/myocardial infarction/chap4");
  source("pde1a.R");
  source("dss004.R");
  source("dss044.R");
```

- The first spatial derivatives

$$\frac{\partial M_{un}(x,t)}{\partial x}, \; \frac{\partial M_1(x,t)}{\partial x}, \; \frac{\partial M_2(x,t)}{\partial x}, \; \frac{\partial IL_{10}(x,t)}{\partial x}, \; \frac{\partial T_\alpha(x,t)}{\partial x}, \; \frac{\partial IL_1(x,t)}{\partial x}$$

are calculated with dss004 using the solution vector returned in matrix out from lsodes. The first subscript of the derivative matrices reflects position x. The second subscript reflects time t.

```
#
# Munx,M1x,M2x,IL10x,Tax,IL1x
  Munx=matrix(0,nrow=nx,ncol=nout);
   M1x=matrix(0,nrow=nx,ncol=nout);
   M2x=matrix(0,nrow=nx,ncol=nout);
 IL10x=matrix(0,nrow=nx,ncol=nout);
   Tax=matrix(0,nrow=nx,ncol=nout);
  IL1x=matrix(0,nrow=nx,ncol=nout);
  for(it in 1:nout){
   Munx[,it]=dss004(xl,xu,nx, Mun[,it]);
    M1x[,it]=dss004(xl,xu,nx,  M1[,it]);
    M2x[,it]=dss004(xl,xu,nx,  M2[,it]);
  IL10x[,it]=dss004(xl,xu,nx,IL10[,it]);
    Tax[,it]=dss004(xl,xu,nx,  Ta[,it]);
   IL1x[,it]=dss004(xl,xu,nx, IL1[,it]);
  }
```

- The homogeneous (zero, no flux) Neumann BCs of eqs. (3.2) are programmed.

```
#
# BCs
  for(it in 1:nout){
    Munx[1,it]=0; Munx[nx,it]=0;
     M1x[1,it]=0;  M1x[nx,it]=0;
     M2x[1,it]=0;  M2x[nx,it]=0;
   IL10x[1,it]=0;IL10x[nx,it]=0;
     Tax[1,it]=0;  Tax[nx,it]=0;
    IL1x[1,it]=0; IL1x[nx,it]=0;
  }
```

Subscripts 1, nx correspond to $x = x_l, x = x_u$, respectively.

- The second spatial derivatives in eqs. (3.1)

$$\frac{\partial^2 M_{un}(x,t)}{\partial x^2}, \ \frac{\partial^2 M_1(x,t)}{\partial x^2}, \ \frac{\partial^2 M_2(x,t)}{\partial x^2}, \ \frac{\partial^2 IL_{10}(x,t)}{\partial x^2}, \ \frac{\partial^2 T_\alpha(x,t)}{\partial x^2}, \ \frac{\partial^2 IL_1(x,t)}{\partial x^2}$$

are calculated with dss044 using the solution vector returned in matrix out from lsodes and the first derivatives calculated previously. As before, the first subscript of the derivative matrices reflects position x, and the second subscript reflects time t.

```
#
# Munxx,M1xx,M2xx,IL10xx,Taxx,IL1xx
   Munxx=matrix(0,nrow=nx,ncol=nout);
    M1xx=matrix(0,nrow=nx,ncol=nout);
    M2xx=matrix(0,nrow=nx,ncol=nout);
  IL10xx=matrix(0,nrow=nx,ncol=nout);
    Taxx=matrix(0,nrow=nx,ncol=nout);
   IL1xx=matrix(0,nrow=nx,ncol=nout);
   nl=2;nu=2;
   for(it in 1:nout){
      Munxx[,it]=dss044(xl,xu,nx, Mun[,it], Munx[,it],nl,nu);
       M1xx[,it]=dss044(xl,xu,nx,  M1[,it],  M1x[,it],nl,nu);
       M2xx[,it]=dss044(xl,xu,nx,  M2[,it],  M2x[,it],nl,nu);
     IL10xx[,it]=dss044(xl,xu,nx,IL10[,it],IL10x[,it],nl,nu);
       Taxx[,it]=dss044(xl,xu,nx,  Ta[,it],  Tax[,it],nl,nu);
      IL1xx[,it]=dss044(xl,xu,nx, IL1[,it], IL1x[,it],nl,nu);
   }
```

nl=2,nu=2 specify Neumann BCs (eqs. (3.2)).
- The t derivative vector,

$$\frac{\partial M_{un}(x,t)}{\partial t}, \ \frac{\partial M_1(x,t)}{\partial t}, \ \frac{\partial M_2(x,t)}{\partial t}, \ \frac{\partial IL_{10}(x,t)}{\partial t}, \ \frac{\partial T_\alpha(x,t)}{\partial t}, \ \frac{\partial IL_1(x,t)}{\partial t}$$

is computed according to the ODE/MOL routine of Listing 3.4, with the index ix for the x interval extended to ix,it for the x and t intervals, respectively.

```
#
# PDE t derivatives
   Munt=matrix(0,nrow=nx,ncol=nout);
    M1t=matrix(0,nrow=nx,ncol=nout);
    M2t=matrix(0,nrow=nx,ncol=nout);
  IL10t=matrix(0,nrow=nx,ncol=nout);
```

```
   Tat=matrix(0,nrow=nx,ncol=nout);
  IL1t=matrix(0,nrow=nx,ncol=nout);
  for(it in 1:nout){
  for(ix in 1:nx){
   fIL1=k2*Mun[ix,it]*IL1[ix,it]/(IL1[ix,it]+cIL1);
    fTa=k3*Mun[ix,it]*Ta[ix,it]/(Ta[ix,it]+cTa);
  fIL10=k4*Mun[ix,it]*IL10[ix,it]/(IL10[ix,it]+cIL10);
    fc1=k5*M2[ix,it]*c1/(c1+IL10[ix,it]);
      fc=c/(c+IL10[ix,it]);
   Munt[ix,it]=DMun*Munxx[ix,it]+M[ix]-fIL1-fTa-fIL10-mu*Mun[ix,it];
    M1t[ix,it]=DM1*M1xx[ix,it]+fIL1+fTa+k1p*M2[ix,it]-k1*M1[ix,it]-
               mu*M1[ix,it];
    M2t[ix,it]=DM2*M2xx[ix,it]+fIL10+k1*M1[ix,it]-k1p*M2[ix,it]-
               mu*M2[ix,it];
  IL10t[ix,it]=DIL10*IL10xx[ix,it]+fc1-dIL10*IL10[ix,it];
    Tat[ix,it]=DTa*Taxx[ix,it]+(k6*M1[ix,it]+lam*Mc[ix])*fc-
               dTa*Ta[ix,it];
   IL1t[ix,it]=DIL1*IL1xx[ix,it]+(k7*M1[ix,it]+lam*Mc[ix])*fc-
               dIL1*IL1[ix,it];
  }
  }
```

- The *t* derivative vectors are plotted in 2D with the R utility matplot. par(mfrow=c(2,2)) specifies a 2×2 matrix of plots on each page of output.

```
#
# Plot PDE t derivatives
# 2D
#
# Mun
# par(mfrow=c(1,1));
  par(mfrow=c(2,2));
  matplot(x=x,y=Munt,type="l",xlab="x",ylab="Munt(x,t)",
          xlim=c(t0,tf),lty=1,main="Munt(x,t)",lwd=2,
          col="black");
#
# M1
  matplot(x=x,y=M1t,type="l",xlab="x",ylab="M1t(x,t)",
          xlim=c(t0,tf),lty=1,main="M1t(x,t)",lwd=2,
```

```
                  col="black");
  #
  # M2
    matplot(x=x,y=M2t,type="l",xlab="x",ylab="M2t(x,t)",
            xlim=c(t0,tf),lty=1,main="M2t(x,t)",lwd=2,
            col="black");
  #
  # IL10
    matplot(x=x,y=IL10t,type="l",xlab="x",ylab="IL10t(x,t)",
            xlim=c(t0,tf),lty=1,main="IL10t(x,t)",lwd=2,
            col="black");
  #
  # Ta
    matplot(x=x,y=Tat,type="l",xlab="x",ylab="Tat(x,t)",
            xlim=c(t0,tf),lty=1,main="Tat(x,t)",lwd=2,
            col="black");
  #
  # IL1
    matplot(x=x,y=IL1t,type="l",xlab="x",ylab="IL1t(x,t)",
            xlim=c(t0,tf),lty=1,main="IL1t(x,t)",lwd=2,
            col="black");
```

- The *t* derivative vectors are plotted in 3D with the R utility persp.

```
  #
  # 3D
    par(mfrow=c(2,2));
    persp(x,tout,Munt,theta=60,phi=45,
          xlim=c(xl,xu),ylim=c(t0,tf),
          xlab="x",ylab="t",zlab="Munt(x,t)");
    persp(x,tout,M1t,theta=60,phi=45,
          xlim=c(xl,xu),ylim=c(t0,tf),
          xlab="x",ylab="t",zlab="M1t(x,t)");
    persp(x,tout,M2t,theta=60,phi=45,
          xlim=c(xl,xu),ylim=c(t0,tf),
          xlab="x",ylab="t",zlab="M2t(x,t)");
    persp(x,tout,IL10t,theta=60,phi=45,
          xlim=c(xl,xu),ylim=c(t0,tf),
          xlab="x",ylab="t",zlab="IL10t(x,t)");
```

```
persp(x,tout,Tat,theta=60,phi=45,
      xlim=c(xl,xu),ylim=c(t0,tf),
      xlab="x",ylab="t",zlab="Tat(x,t)");
persp(x,tout,IL1t,theta=60,phi=45,
      xlim=c(xl,xu),ylim=c(t0,tf),
      xlab="x",ylab="t",zlab="IL1t(x,t)");
```

This completes the calculation and display of the *t* derivatives of eq. (3.1).

4.1.2 ODE/MOL routine

The ODE/MOL routine is in Listing 3.4.

4.1.3 Numerical, graphical output

The output from `nrow,ncol` called in the main program of Listing 3.3 (dimensions of the solution matrix `out` from `lsodes`) follows.

```
[1]  11
```

```
[1]  247
```

The row dimension of the solution matrix `out` from `lsodes` is 11 corresponding to the definition of the *t* values of `tout`.

```
#
# Independent variable for ODE integration
  t0=0;tf=1;nout=11;
  tout=seq(from=t0,to=tf,by=(tf-t0)/(nout-1));
```

The column dimension of `out` is 247 corresponding to the 246 ODE dependent variables $M_{un}(x,t)$ to $IL_1(x,t)$ of eqs. (3.1) and the independent variable *t* for the 246 dependent variables.

For **ncase=1**, the graphical output of Figs. 4.1-1,2,3,4 indicates that the six dependent variables are invariant in *x* (in accordance with Figs. 3.5-1,2,3,4), and Figs. 4.1-5,6,7,8 indicate the six derivatives in *t* are invariant in *x*.

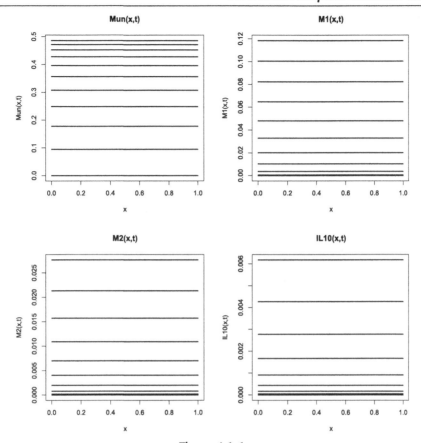

Figure 4.1-1:
Numerical $M_{nu}(x,t)$, $M_1(x,t)$, $M_2(x,t)$, $IL_{10}(x,t)$, from eqs. (3.1), (3.2), (3.3), Listings 3.3, 4.1 `ncase=1`, 2D

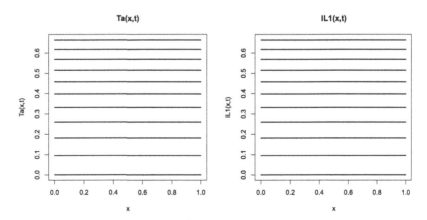

Figure 4.1-2:
Numerical $T\alpha(x,t)$, $IL_1(x,t)$, from eqs. (3.1), (3.2), (3.3), Listings 3.3, 4.1 `ncase=1`, 2D

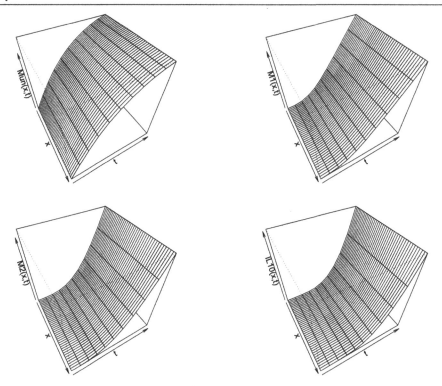

Figure 4.1-3:
Numerical $M_{nu}(x,t)$, $M_1(x,t)$, $M_2(x,t)$, $IL_{10}(x,t)$, from eqs. (3.1), (3.2), (3.3), Listings 3.3, 4.1 ncase=1, 3D

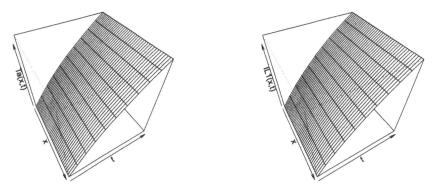

Figure 4.1-4:
Numerical $T\alpha(x,t)$, $IL_1(x,t)$, from eqs. (3.1), (3.2), (3.3), Listings 3.3, 4.1 ncase=1, 3D

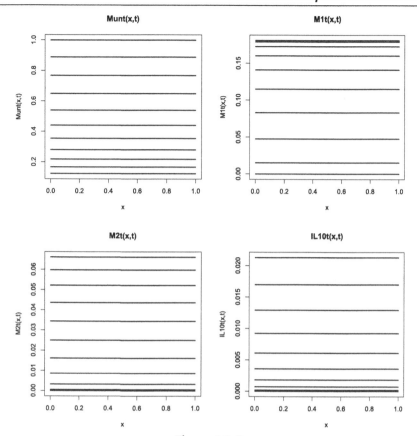

Figure 4.1-5:

Numerical $\dfrac{\partial M_{nu}(x,t)}{\partial t}$, $\dfrac{\partial M_1(x,t)}{\partial t}$, $\dfrac{\partial M_2(x,t)}{\partial t}$, $\dfrac{\partial IL_{10}(x,t)}{\partial t}$, from eqs. (3.1), (3.2), (3.3), Listings 3.3, 4.1 ncase=1, 2D

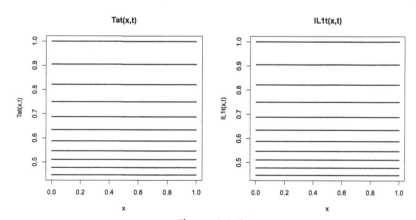

Figure 4.1-6:

Numerical $\dfrac{\partial T\alpha(x,t)}{\partial t}$, $\dfrac{\partial IL_1(x,t)}{\partial t}$, from eqs. (3.1), (3.2), (3.3), Listings 3.3, 4.1 ncase=1, 2D

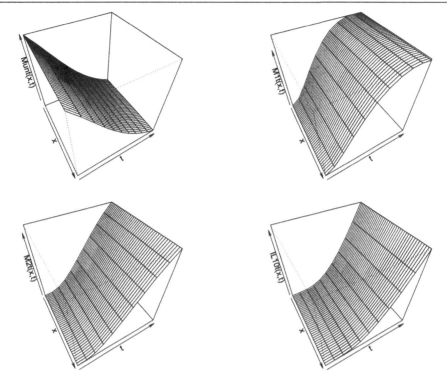

Figure 4.1-7:

Numerical $\dfrac{\partial M_{nu}(x,t)}{\partial t}$, $\dfrac{\partial M_1(x,t)}{\partial t}$, $\dfrac{\partial M_2(x,t)}{\partial t}$, $\dfrac{\partial IL_{10}(x,t)}{\partial t}$, from eqs. (3.1), (3.2), (3.3), Listings 3.3, 4.1 ncase=1, 3D

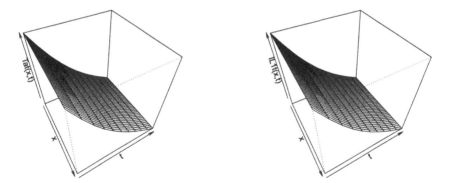

Figure 4.1-8:

Numerical $\dfrac{\partial T\alpha(x,t)}{\partial t}$, $\dfrac{\partial IL_1(x,t)}{\partial t}$, from eqs. (3.1), (3.2), (3.3), Listings 3.3, 4.1 ncase=1, 3D

For **ncase=2**, the graphical output of Figs. 4.2-1,2,3,4 indicates that the six dependent variable derivatives in t are variant in x, in accordance with Figs. 3.5-5,6,7,8.

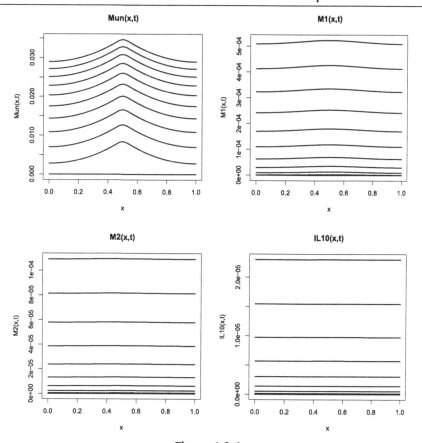

Figure 4.2-1:
Numerical $M_{nu}(x,t)$, $M_1(x,t)$, $M_2(x,t)$, $IL_{10}(x,t)$, from eqs. (3.1), (3.2), (3.3), Listings 3.3, 4.1
`ncase=2, 2D`

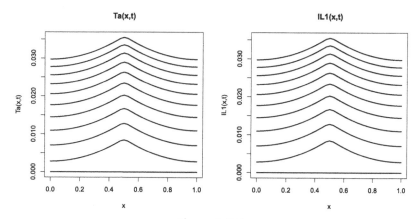

Figure 4.2-2:
Numerical $T\alpha(x,t)$, $IL_1(x,t)$, from eqs. (3.1), (3.2), (3.3), Listings 3.3, 4.1 `ncase=2, 2D`

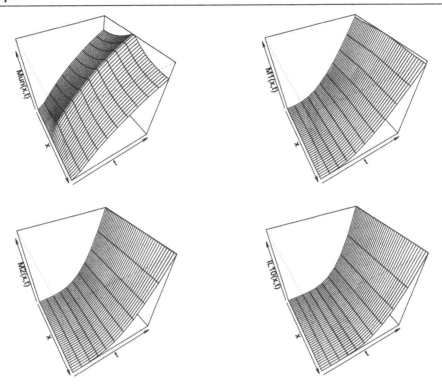

Figure 4.2-3:
Numerical $M_{nu}(x,t)$, $M_1(x,t)$, $M_2(x,t)$, $IL_{10}(x,t)$, from eqs. (3.1), (3.2), (3.3), Listings 3.3, 4.1
ncase=2, 3D

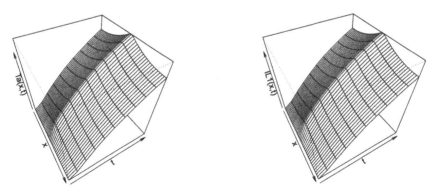

Figure 4.2-4:
Numerical $T\alpha(x,t)$, $IL_1(x,t)$, from eqs. (3.1), (3.2), (3.3), Listings 3.3, 4.1 ncase=2, 3D

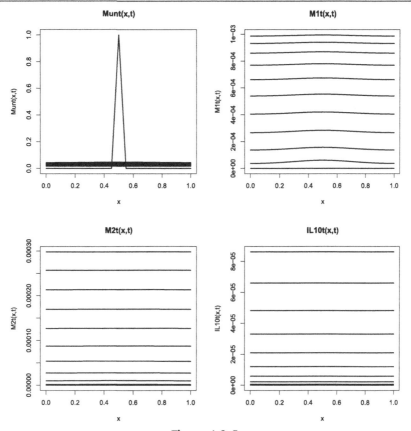

Figure 4.2-5:
Numerical $\dfrac{\partial M_{nu}(x,t)}{\partial t}$, $\dfrac{\partial M_1(x,t)}{\partial t}$, $\dfrac{\partial M_2(x,t)}{\partial t}$, $\dfrac{\partial IL_{10}(x,t)}{\partial t}$, from eqs. (3.1), (3.2), (3.3),
Listings 3.3, 4.1 ncase=2, 2D

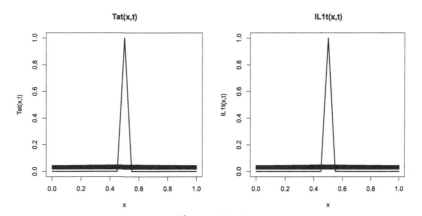

Figure 4.2-6:
Numerical $\dfrac{\partial T\alpha(x,t)}{\partial t}$, $\dfrac{\partial IL_1(x,t)}{\partial t}$, from eqs. (3.1), (3.2), (3.3), Listings 3.3, 4.1 ncase=2, 2D

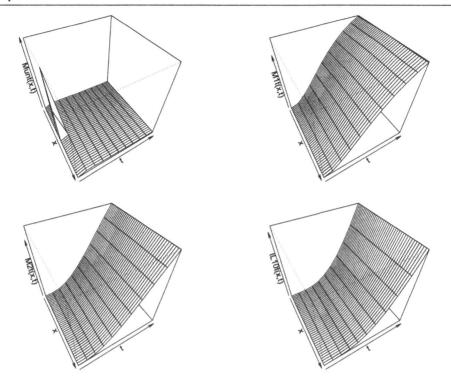

Figure 4.2-7:
Numerical $\dfrac{\partial M_{nu}(x,t)}{\partial t}$, $\dfrac{\partial M_1(x,t)}{\partial t}$, $\dfrac{\partial M_2(x,t)}{\partial t}$, $\dfrac{\partial IL_{10}(x,t)}{\partial t}$, from eqs. (3.1), (3.2), (3.3),
Listings 3.3, 4.1 ncase=2, 3D

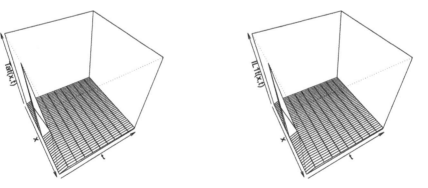

Figure 4.2-8:
Numerical $\dfrac{\partial T\alpha(x,t)}{\partial t}$, $\dfrac{\partial IL_1(x,t)}{\partial t}$, from eqs. (3.1), (3.2), (3.3), Listings 3.3, 4.1 ncase=2, 3D

The x-variation of the t derivatives for `ncase=2` is clear, particularly for $\dfrac{\partial M_{un}(x,t)}{\partial t}$, $\dfrac{\partial T\alpha(x,t)}{\partial t}$, $\dfrac{\partial IL_1(x,t)(x,t)}{\partial t}$ resulting from the $M(x,t)$, $M_c(x,t)$ (triangular pulses) of Listing 3.3.

Also, the solutions do not reach a steady state because of the continuing effect in t of the pulses in Listing 3.3.

The response to the triangular pulses of Listing 3.3 is a nearly instantaneous smoothing by the diffusion terms in eqs. (3.1). This smoothing can be investigated by varying the diffusivities. For example, if

```
#
# Diffusivities
   DMun=1;
    DM1=1;
    DM2=1;
  DIL10=1;
    DTa=1;
   DIL1=1;
```

in Listing 3.3 is changed to

```
#
# Diffusivities
   DMun=0.1;
    DM1=0.1;
    DM2=0.1;
  DIL10=0.1;
    DTa=0.1;
   DIL1=0.1;
```

the graphical output in Figs. 4.2-1 to 4.2-8 indicates a reduced diffusive smoothing of the solution.

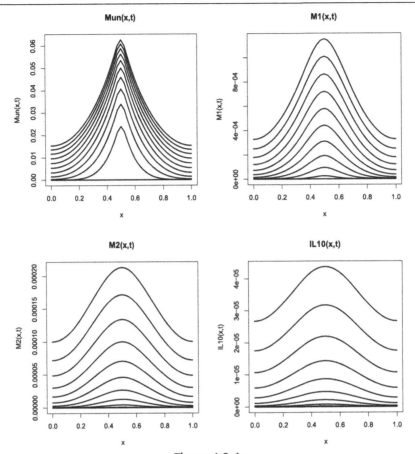

Figure 4.3-1:
Numerical $M_{nu}(x,t)$, $M_1(x,t)$, $M_2(x,t)$, $L_{10}(x,t)$, from eqs. (3.1), (3.2), (3.3), reduced diffusivities, ncase=2, 2D

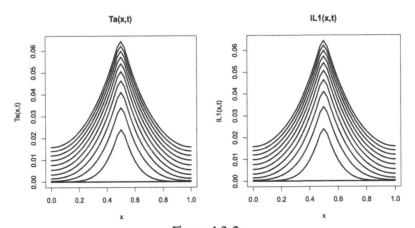

Figure 4.3-2:
Numerical $T\alpha(x,t)$, $IL_1(x,t)$, from eqs. (3.1), (3.2), (3.3), reduced diffusivities, ncase=2, 2D

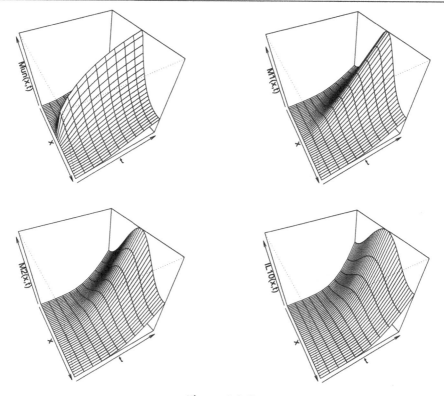

Figure 4.3-3:
Numerical $M_{nu}(x,t)$, $M_1(x,t)$, $M_2(x,t)$, $L_{10}(x,t)$, from eqs. (3.1), (3.2), (3.3), reduced diffusivities, ncase=2, 3D

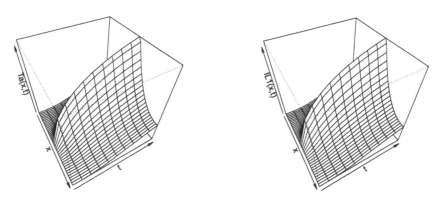

Figure 4.3-4:
Numerical $T\alpha(x,t)$, $IL_1(x,t)$, from eqs. (3.1), (3.2), (3.3), reduced diffusivities, ncase=2, 3D

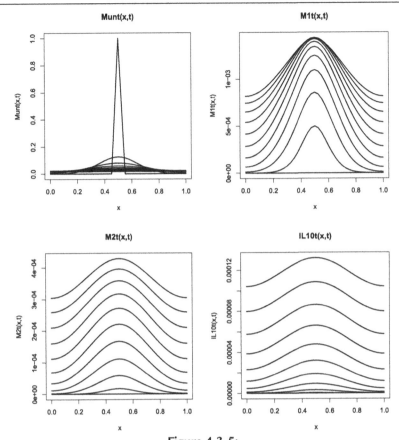

Figure 4.3-5:

Numerical $\dfrac{\partial M_{nu}(x,t)}{\partial t}$, $\dfrac{\partial M_1(x,t)}{\partial t}$, $\dfrac{\partial M_2(x,t)}{\partial t}$, $\dfrac{\partial IL_{10}(x,t)}{\partial t}$, from eqs. (3.1), (3.2), (3.3), reduced diffusivities, ncase=2, 2D

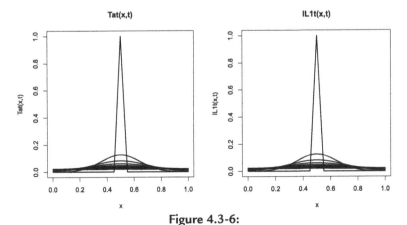

Figure 4.3-6:

Numerical $\dfrac{\partial T\alpha(x,t)}{\partial t}$, $\dfrac{\partial IL_1(x,t)}{\partial t}$, from eqs. (3.1), (3.2), (3.3), reduced diffusivities, ncase=2, 2D

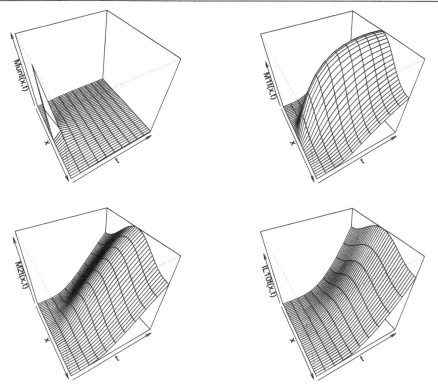

Figure 4.3-7:

Numerical $\dfrac{\partial M_{nu}(x,t)}{\partial t}$, $\dfrac{\partial M_1(x,t)}{\partial t}$, $\dfrac{\partial M_2(x,t)}{\partial t}$, $\dfrac{\partial IL_{10}(x,t)}{\partial t}$, from eqs. (3.1), (3.2), (3.3), reduced diffusivities, ncase=2, 3D

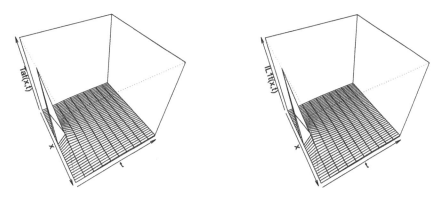

Figure 4.3-8:

Numerical $\dfrac{\partial T\alpha(x,t)}{\partial t}$, $\dfrac{\partial IL_1(x,t)}{\partial t}$, from eqs. (3.1), (3.2), (3.3), reduced diffusivities, ncase=2, 3D

In summary, a comparison of Figs. 4.2 and Figs. 4.3 indicates the diffusivities in eqs. (3.1) are sensitive parameters in determining the properties (smoothing) of the solutions $M_{un}(x, t)$, $M_1(x, t)$, $M_2(x, t)$, $IL_{10}(x, t)$, $T_\alpha(x, t)$, $IL_1(x, t)$.

4.2 Summary and conclusions

The analysis of the six PDE model of eqs. (3.1) is extended to include the LHS derivatives in t, which are computed and displayed as a function of x and t. Three cases are considered: (1) $M(x, t)$, $M_c(x, t)$ constant (output in Figs. 4.1), (2) $M(x, t)$, $M_c(x, t)$ variable in x, (output in Figs. 4.2), and (3) reduced diffusivities $D_{M_{un}}$ to D_{IL_1} in eqs. (3.1) (output in Figs. 4.3). The t derivatives define the features of the solutions of the six PDE dependent variables $M_{un}(x, t)$, $M_1(x, t)$, $M_2(x, t)$, $IL_{10}(x, t)$, $T_\alpha(x, t)$, $IL_1(x, t)$ according to eqs. (3.1).

In the next chapter, the RHS terms of eqs. (3.1) are computed and displayed.

References

[1] K. Soetaert, J. Cash, F. Mazzia, Solving Differential Equations in R, Springer-Verlag, Heidelberg, Germany, 2012.

## CHAPTER 5

Analysis of the PDE model terms

Introduction

The features of the PDE solutions of eqs. (3.1), (3.2), (3.3), e.g., in Figs. 3.5, are determined by the left hand side (LHS) derivatives in t of eqs. (3.1). These derivatives in turn are determined by the RHS terms of eqs. (3.1). The PDE RHS terms are computed and displayed in this chapter for additional insight into the PDE modeling of post-MI [1].

5.1 RHS terms of the PDE model

The following calculation and display of the PDE RHS terms gives a detailed explanation of the features of the eqs. (3.1) solutions.

5.1.1 Main program extension

The following coding for the calculation and display of the PDE RHS terms is added to the end of Listing 4.1 (with diffusivities equal to 0.1).

```
#
# Six PDE post-MI model
# RHS terms
#
# Delete previous workspaces
  rm(list=ls(all=TRUE))
#
# Access ODE integrator
  library("deSolve");
#
# Access functions for numerical solution
  setwd("g:/myocardial infarction/chap5");
  source("pde1a.R");
  source("dss004.R");
  source("dss044.R");
```

Modeling of Post-Myocardial Infarction
https://doi.org/10.1016/B978-0-44-313611-5.00010-X

```
                .                 .

            .               .

            .               .

#
# PDE LHS, RHS terms
#
# Mun(x,t)
  term11=matrix(0,nrow=nx,ncol=nout);
  term12=matrix(0,nrow=nx,ncol=nout);
  term13=matrix(0,nrow=nx,ncol=nout);
  term14=matrix(0,nrow=nx,ncol=nout);
  term15=matrix(0,nrow=nx,ncol=nout);
  term16=matrix(0,nrow=nx,ncol=nout);
   pde1t=matrix(0,nrow=nx,ncol=nout);
  for(it in 1:nout){
  for(ix in 1:nx){
   fIL1=k2*Mun[ix,it]*IL1[ix,it]/(IL1[ix,it]+cIL1);
     fTa=k3*Mun[ix,it]*Ta[ix,it]/(Ta[ix,it]+cTa);
  fIL10=k4*Mun[ix,it]*IL10[ix,it]/(IL10[ix,it]+cIL10);
  term11[ix,it]=DMun*Munxx[ix,it];
  term12[ix,it]=M[ix];
  term13[ix,it]=-fIL1;
  term14[ix,it]=-fTa;
  term15[ix,it]=-fIL10;
  term16[ix,it]=-mu*Mun[ix,it];
   pde1t[ix,it]=term11[ix,it]+term12[ix,it]+term13[ix,it]+
                 term14[ix,it]+term15[ix,it]+term16[ix,it];
  }
  }
#
# Plot Mun(x,t) RHS, LHS terms
#
# term11
  par(mfrow=c(2,2));
  matplot(x=x,y=term11,type="l",xlab="x",ylab="term11(x,t)",
          xlim=c(xl,xu),lty=1,main="term11(x,t)",lwd=2,
          col="black");
#
# term12
```

```
  matplot(x=x,y=term12,type="l",xlab="x",ylab="term12(x,t)",
          xlim=c(xl,xu),lty=1,main="term12(x,t)",lwd=2,
          col="black");
#
# term13
  matplot(x=x,y=term13,type="l",xlab="x",ylab="term13(x,t)",
          xlim=c(xl,xu),lty=1,main="term13(x,t)",lwd=2,
          col="black");
#
# term14
  matplot(x=x,y=term14,type="l",xlab="x",ylab="term14(x,t)",
          xlim=c(xl,xu),lty=1,main="term14(x,t)",lwd=2,
          col="black");
#
# term15
  matplot(x=x,y=term15,type="l",xlab="x",ylab="term15(x,t)",
          xlim=c(xl,xu),lty=1,main="term15(x,t)",lwd=2,
          col="black");
#
# term16
  matplot(x=x,y=term16,type="l",xlab="x",ylab="term16(x,t)",
          xlim=c(xl,xu),lty=1,main="term16(x,t)",lwd=2,
          col="black");
#
# pde1t
  matplot(x=x,y=pde1t,type="l",xlab="x",ylab="pde1t(x,t)",
          xlim=c(xl,xu),lty=1,main="pde1t(x,t)",lwd=2,
          col="black");
#
# M1(x,t)
  term21=matrix(0,nrow=nx,ncol=nout);
  term22=matrix(0,nrow=nx,ncol=nout);
  term23=matrix(0,nrow=nx,ncol=nout);
  term24=matrix(0,nrow=nx,ncol=nout);
  term25=matrix(0,nrow=nx,ncol=nout);
  term26=matrix(0,nrow=nx,ncol=nout);
   pde2t=matrix(0,nrow=nx,ncol=nout);
  for(it in 1:nout){
  for(ix in 1:nx){
```

```
    fIL1=k2*Mun[ix,it]*IL1[ix,it]/(IL1[ix,it]+cIL1);
     fTa=k3*Mun[ix,it]*Ta[ix,it]/(Ta[ix,it]+cTa);
  term21[ix,it]=DM1*M1xx[ix,it];
  term22[ix,it]=fIL1;
  term23[ix,it]=fTa;
  term24[ix,it]=k1p*M2[ix,it];
  term25[ix,it]=-k1*M1[ix,it];
  term26[ix,it]=-mu*M1[ix,it];
   pde2t[ix,it]=term21[ix,it]+term22[ix,it]+term23[ix,it]+
                 term24[ix,it]+term25[ix,it]+term26[ix,it];
  }
  }
#
# Plot M1(x,t) RHS, LHS terms
#
# term21
  par(mfrow=c(2,2));
  matplot(x=x,y=term21,type="l",xlab="x",ylab="term21(x,t)",
          xlim=c(xl,xu),lty=1,main="term21(x,t)",lwd=2,
          col="black");
#
# term22
  matplot(x=x,y=term22,type="l",xlab="x",ylab="term22(x,t)",
          xlim=c(xl,xu),lty=1,main="term22(x,t)",lwd=2,
          col="black");
#
# term23
  matplot(x=x,y=term23,type="l",xlab="x",ylab="term23(x,t)",
          xlim=c(xl,xu),lty=1,main="term23(x,t)",lwd=2,
          col="black");
#
# term24
  matplot(x=x,y=term24,type="l",xlab="x",ylab="term24(x,t)",
          xlim=c(xl,xu),lty=1,main="term24(x,t)",lwd=2,
          col="black");
#
# term25
  matplot(x=x,y=term25,type="l",xlab="x",ylab="term25(x,t)",
          xlim=c(xl,xu),lty=1,main="term25(x,t)",lwd=2,
```

```
                 col="black");
#
# term26
  matplot(x=x,y=term26,type="l",xlab="x",ylab="term26(x,t)",
          xlim=c(xl,xu),lty=1,main="term26(x,t)",lwd=2,
          col="black");
#
# pde2t
  matplot(x=x,y=pde2t,type="l",xlab="x",ylab="pde2t(x,t)",
          xlim=c(xl,xu),lty=1,main="pde2t(x,t)",lwd=2,
          col="black");
#
# M2(x,t)
  term31=matrix(0,nrow=nx,ncol=nout);
  term32=matrix(0,nrow=nx,ncol=nout);
  term33=matrix(0,nrow=nx,ncol=nout);
  term34=matrix(0,nrow=nx,ncol=nout);
  term35=matrix(0,nrow=nx,ncol=nout);
   pde3t=matrix(0,nrow=nx,ncol=nout);
  for(it in 1:nout){
  for(ix in 1:nx){
  fIL10=k4*Mun[ix,it]*IL10[ix,it]/(IL10[ix,it]+cIL10);
  term31[ix,it]=DM2*M2xx[ix,it];
  term32[ix,it]=fIL10;
  term33[ix,it]=k1*M1[ix,it];
  term34[ix,it]=-k1p*M2[ix,it];
  term35[ix,it]=-mu*M2[ix,it];
   pde3t[ix,it]=term31[ix,it]+term32[ix,it]+term33[ix,it]+
                term34[ix,it]+term35[ix,it];
  }
  }
#
# Plot M2(x,t) RHS, LHS terms
#
# term31
  par(mfrow=c(2,2));
  matplot(x=x,y=term31,type="l",xlab="x",ylab="term31(x,t)",
          xlim=c(xl,xu),lty=1,main="term31(x,t)",lwd=2,
          col="black");
```

```
#
# term32
  matplot(x=x,y=term32,type="l",xlab="x",ylab="term32(x,t)",
          xlim=c(xl,xu),lty=1,main="term32(x,t)",lwd=2,
          col="black");
#
# term33
  matplot(x=x,y=term33,type="l",xlab="x",ylab="term33(x,t)",
          xlim=c(xl,xu),lty=1,main="term33(x,t)",lwd=2,
          col="black");
#
# term34
  matplot(x=x,y=term34,type="l",xlab="x",ylab="term34(x,t)",
          xlim=c(xl,xu),lty=1,main="term34(x,t)",lwd=2,
          col="black");
#
# term35
  matplot(x=x,y=term35,type="l",xlab="x",ylab="term35(x,t)",
          xlim=c(xl,xu),lty=1,main="term35(x,t)",lwd=2,
          col="black");
# pde3t
  matplot(x=x,y=pde3t,type="l",xlab="x",ylab="pde3t(x,t)",
          xlim=c(xl,xu),lty=1,main="pde3t(x,t)",lwd=2,
          col="black");
#
# IL10(x,t)
  term41=matrix(0,nrow=nx,ncol=nout);
  term42=matrix(0,nrow=nx,ncol=nout);
  term43=matrix(0,nrow=nx,ncol=nout);
   pde4t=matrix(0,nrow=nx,ncol=nout);
  for(it in 1:nout){
  for(ix in 1:nx){
  fc1=k5*M2[ix,it]*c1/(c1+IL10[ix,it]);
  term41[ix,it]=DIL10*IL10xx[ix,it];
  term42[ix,it]=fc1;
  term43[ix,it]=-dIL10*IL10[ix,it];
   pde4t[ix,it]=term41[ix,it]+term42[ix,it]+term43[ix,it];
  }
  }
```

```
#
# Plot IL10(x,t) RHS, LHS terms
#
# term41
  par(mfrow=c(2,2));
  matplot(x=x,y=term41,type="l",xlab="x",ylab="term41(x,t)",
          xlim=c(xl,xu),lty=1,main="term41(x,t)",lwd=2,
          col="black");
#
# term42
  matplot(x=x,y=term42,type="l",xlab="x",ylab="term42(x,t)",
          xlim=c(xl,xu),lty=1,main="term42(x,t)",lwd=2,
          col="black");
#
# term43
  matplot(x=x,y=term43,type="l",xlab="x",ylab="term43(x,t)",
          xlim=c(xl,xu),lty=1,main="term43(x,t)",lwd=2,
          col="black");
# pde4t
  matplot(x=x,y=pde4t,type="l",xlab="x",ylab="pde4t(x,t)",
          xlim=c(xl,xu),lty=1,main="pde4t(x,t)",lwd=2,
          col="black");
#
# Ta(x,t)
  term51=matrix(0,nrow=nx,ncol=nout);
  term52=matrix(0,nrow=nx,ncol=nout);
  term53=matrix(0,nrow=nx,ncol=nout);
  pde5t=matrix(0,nrow=nx,ncol=nout);
  for(it in 1:nout){
  for(ix in 1:nx){
  fc=c/(c+IL10[ix,it]);
  term51[ix,it]=DTa*Taxx[ix,it];
  term52[ix,it]=(k6*M1[ix,it]+lam*Mc[ix])*fc;
  term53[ix,it]=-dTa*Ta[ix,it];
  pde5t[ix,it]=term51[ix,it]+term52[ix,it]+term53[ix,it];
  }
  }
#
# Plot Ta(x,t) RHS, LHS terms
```

```
#
# term51
  par(mfrow=c(2,2));
  matplot(x=x,y=term51,type="l",xlab="x",ylab="term51(x,t)",
          xlim=c(xl,xu),lty=1,main="term51(x,t)",lwd=2,
          col="black");
#
# term52
  matplot(x=x,y=term52,type="l",xlab="x",ylab="term52(x,t)",
          xlim=c(xl,xu),lty=1,main="term52(x,t)",lwd=2,
          col="black");
#
# term53
  matplot(x=x,y=term53,type="l",xlab="x",ylab="term53(x,t)",
          xlim=c(xl,xu),lty=1,main="term53(x,t)",lwd=2,
          col="black");
#
# pde5t
  matplot(x=x,y=pde5t,type="l",xlab="x",ylab="pde5t(x,t)",
          xlim=c(xl,xu),lty=1,main="pde5t(x,t)",lwd=2,
          col="black");
#
# IL1(x,t)
  term61=matrix(0,nrow=nx,ncol=nout);
  term62=matrix(0,nrow=nx,ncol=nout);
  term63=matrix(0,nrow=nx,ncol=nout);
  pde6t=matrix(0,nrow=nx,ncol=nout);
  for(it in 1:nout){
  for(ix in 1:nx){
  fc=c/(c+IL10[ix,it]);
  term61[ix,it]=DIL1*IL1xx[ix,it];
  term62[ix,it]=(k7*M1[ix,it]+lam*Mc[ix])*fc;
  term63[ix,it]=-dIL1*IL1[ix,it];
  pde6t[ix,it]=term61[ix,it]+term62[ix,it]+term63[ix,it];
  }
  }
#
# Plot IL1(x,t) RHS, LHS terms
#
```

```
# term61
  par(mfrow=c(2,2));
  matplot(x=x,y=term61,type="l",xlab="x",ylab="term61(x,t)",
          xlim=c(xl,xu),lty=1,main="term61(x,t)",lwd=2,
          col="black");
#
# term62
  matplot(x=x,y=term62,type="l",xlab="x",ylab="term62(x,t)",
          xlim=c(xl,xu),lty=1,main="term62(x,t)",lwd=2,
          col="black");
#
# term63
  matplot(x=x,y=term63,type="l",xlab="x",ylab="term63(x,t)",
          xlim=c(xl,xu),lty=1,main="term63(x,t)",lwd=2,
          col="black");
#
# pde6t
  matplot(x=x,y=pde1t,type="l",xlab="x",ylab="pde6t(x,t)",
          xlim=c(xl,xu),lty=1,main="pde6t(x,t)",lwd=2,
          col="black");
```

Listing 5.1: Addition to Listing 4.1 for the calculation and display of the PDE RHS, LHS terms

Listing 5.1 requires some additional explanation.

The following coding is for the RHS and LHS of eq. (3.1-1).

•   Matrices for the six RHS terms of eq. (3.1-1) and the LHS t derivative are defined.

```
#
# PDE LHS, RHS terms
#
# Mun(x,t)
  term11=matrix(0,nrow=nx,ncol=nout);
  term12=matrix(0,nrow=nx,ncol=nout);
  term13=matrix(0,nrow=nx,ncol=nout);
  term14=matrix(0,nrow=nx,ncol=nout);
  term15=matrix(0,nrow=nx,ncol=nout);
  term16=matrix(0,nrow=nx,ncol=nout);
   pde1t=matrix(0,nrow=nx,ncol=nout);
```

The terms are named with two numbers. The first number designates the PDE. The second number designates the term.

For example, `term11` is the first term in the first PDE (eq. (3.1-1)).

- The six RHS terms, `term11` to `term16`, and the LHS term, `pde1t` are computed over x with an inner `for` and index `ix`, and over t with an outer `for` and index `it`.

```
for(it in 1:nout){
for(ix in 1:nx){
 fIL1=k2*Mun[ix,it]*IL1[ix,it]/(IL1[ix,it]+cIL1);
  fTa=k3*Mun[ix,it]*Ta[ix,it]/(Ta[ix,it]+cTa);
fIL10=k4*Mun[ix,it]*IL10[ix,it]/(IL10[ix,it]+cIL10);
term11[ix,it]=DMun*Munxx[ix,it];
term12[ix,it]=M[ix];
term13[ix,it]=-fIL1;
term14[ix,it]=-fTa;
term15[ix,it]=-fIL10;
term16[ix,it]=-mu*Mun[ix,it];
 pde1t[ix,it]=term11[ix,it]+term12[ix,it]+term13[ix,it]+
              term14[ix,it]+term15[ix,it]+term16[ix,it];
}
}
```

- The six RHS terms and LHS t derivative are plotted in 2D with the R utility `matplot`.

```
#
# Plot Mun(x,t) RHS, LHS terms
#
# term11
  par(mfrow=c(2,2));
  matplot(x=x,y=term11,type="l",xlab="x",ylab="term11(x,t)",
          xlim=c(xl,xu),lty=1,main="term11(x,t)",lwd=2,
          col="black");
#
# term12
  matplot(x=x,y=term12,type="l",xlab="x",ylab="term12(x,t)",
          xlim=c(xl,xu),lty=1,main="term12(x,t)",lwd=2,
          col="black");
#
# term13
  matplot(x=x,y=term13,type="l",xlab="x",ylab="term13(x,t)",
```

```
            xlim=c(xl,xu),lty=1,main="term13(x,t)",lwd=2,
            col="black");
#
# term14
  matplot(x=x,y=term14,type="l",xlab="x",ylab="term14(x,t)",
            xlim=c(xl,xu),lty=1,main="term14(x,t)",lwd=2,
            col="black");
#
# term15
  matplot(x=x,y=term15,type="l",xlab="x",ylab="term15(x,t)",
            xlim=c(xl,xu),lty=1,main="term15(x,t)",lwd=2,
            col="black");
#
# term16
  matplot(x=x,y=term16,type="l",xlab="x",ylab="term16(x,t)",
            xlim=c(xl,xu),lty=1,main="term16(x,t)",lwd=2,
            col="black");
#
# pde1t
  matplot(x=x,y=pde1t,type="l",xlab="x",ylab="pde1t(x,t)",
            xlim=c(xl,xu),lty=1,main="pde1t(x,t)",lwd=2,
            col="black");
```

The plots are parametric in t (see Figs. 5.1-1,2). Default scaling of the y (*ordinate*) variable is used in each call to `matplot`.
`par(mfrow=c(2,2))` specifies a 2×2 matrix of plots on each page.

The following coding is for eq. (3.1-2), with some repetition of the preceding discussion for eq. (3.1-1) so that the explanation is self contained.

- Matrices for the six RHS terms of eq. (3.1-2) and the LHS t derivative are defined.

```
#
# M1(x,t)
  term21=matrix(0,nrow=nx,ncol=nout);
  term22=matrix(0,nrow=nx,ncol=nout);
  term23=matrix(0,nrow=nx,ncol=nout);
  term24=matrix(0,nrow=nx,ncol=nout);
  term25=matrix(0,nrow=nx,ncol=nout);
  term26=matrix(0,nrow=nx,ncol=nout);
   pde2t=matrix(0,nrow=nx,ncol=nout);
```

The terms are named with two numbers. The first number designates the PDE. The second number designates the term.

For example, `term21` is the first term in the second PDE (eq. (3.1-2)).

- The six RHS terms, `term21` to `term26`, and the LHS term, `pde2t` are computed over x with an inner `for` and index `ix`, and over t with an outer `for` and index `it`.

```
for(it in 1:nout){
for(ix in 1:nx){
 fIL1=k2*Mun[ix,it]*IL1[ix,it]/(IL1[ix,it]+cIL1);
  fTa=k3*Mun[ix,it]*Ta[ix,it]/(Ta[ix,it]+cTa);
term21[ix,it]=DM1*M1xx[ix,it];
term22[ix,it]=fIL1;
term23[ix,it]=fTa;
term24[ix,it]=k1p*M2[ix,it];
term25[ix,it]=-k1*M1[ix,it];
term26[ix,it]=-mu*M1[ix,it];
 pde2t[ix,it]=term21[ix,it]+term22[ix,it]+term23[ix,it]+
              term24[ix,it]+term25[ix,it]+term26[ix,it];
}
}
```

- The six RHS terms and LHS t derivative are plotted in 2D with the R utility `matplot`.

```
#
# Plot M1(x,t) RHS, LHS terms
#
# term21
  par(mfrow=c(2,2));
  matplot(x=x,y=term21,type="l",xlab="x",ylab="term21(x,t)",
          xlim=c(xl,xu),lty=1,main="term21(x,t)",lwd=2,
          col="black");
#
# term22
  matplot(x=x,y=term22,type="l",xlab="x",ylab="term22(x,t)",
          xlim=c(xl,xu),lty=1,main="term22(x,t)",lwd=2,
          col="black");
#
# term23
  matplot(x=x,y=term23,type="l",xlab="x",ylab="term23(x,t)",
          xlim=c(xl,xu),lty=1,main="term23(x,t)",lwd=2,
```

```
                      col="black");
  #
  # term24
    matplot(x=x,y=term24,type="l",xlab="x",ylab="term24(x,t)",
            xlim=c(xl,xu),lty=1,main="term24(x,t)",lwd=2,
            col="black");
  #
  # term25
    matplot(x=x,y=term25,type="l",xlab="x",ylab="term25(x,t)",
            xlim=c(xl,xu),lty=1,main="term25(x,t)",lwd=2,
            col="black");
  #
  # term26
    matplot(x=x,y=term26,type="l",xlab="x",ylab="term26(x,t)",
            xlim=c(xl,xu),lty=1,main="term26(x,t)",lwd=2,
            col="black");
  #
  # pde2t
    matplot(x=x,y=pde2t,type="l",xlab="x",ylab="pde2t(x,t)",
            xlim=c(xl,xu),lty=1,main="pde2t(x,t)",lwd=2,
            col="black");
```

The plots are parametric in t (see Figs. 5.2-1,2). Default scaling of the y (*ordinate*) variable is used in each call to `matplot`.

`par(mfrow=c(2,2))` specifies a 2×2 matrix of plots on each page.

The following coding is for eq. (3.1-3), with some repetition of the preceding discussion for eqs. (3.1-1,2) so that the explanation is self contained.

• Matrices for the five RHS terms of eq. (3.1-3) and the LHS t derivative are defined.

```
  #
  # M2(x,t)
    term31=matrix(0,nrow=nx,ncol=nout);
    term32=matrix(0,nrow=nx,ncol=nout);
    term33=matrix(0,nrow=nx,ncol=nout);
    term34=matrix(0,nrow=nx,ncol=nout);
    term35=matrix(0,nrow=nx,ncol=nout);
     pde3t=matrix(0,nrow=nx,ncol=nout);
```

The terms are named with two numbers. The first number designates the PDE. The second number designates the term.

For example, term31 is the first term in the third PDE (eq. (3.1-3)).

- The five RHS terms, term31 to term35, and the LHS term, pde3t are computed over x with an inner for and index ix, and over t with an outer for and index it.

```
for(it in 1:nout){
for(ix in 1:nx){
fIL10=k4*Mun[ix,it]*IL10[ix,it]/(IL10[ix,it]+cIL10);
term31[ix,it]=DM2*M2xx[ix,it];
term32[ix,it]=fIL10;
term33[ix,it]=k1*M1[ix,it];
term34[ix,it]=-k1p*M2[ix,it];
term35[ix,it]=-mu*M2[ix,it];
  pde3t[ix,it]=term31[ix,it]+term32[ix,it]+term33[ix,it]+
              term34[ix,it]+term35[ix,it];
}
}
```

- The five RHS terms and LHS t derivative are plotted in 2D with the R utility matplot.

```
#
# Plot M2(x,t) RHS, LHS terms
#
# term31
  par(mfrow=c(2,2));
  matplot(x=x,y=term31,type="l",xlab="x",ylab="term31(x,t)",
          xlim=c(xl,xu),lty=1,main="term31(x,t)",lwd=2,
          col="black");
#
# term32
  matplot(x=x,y=term32,type="l",xlab="x",ylab="term32(x,t)",
          xlim=c(xl,xu),lty=1,main="term32(x,t)",lwd=2,
          col="black");
#
# term33
  matplot(x=x,y=term33,type="l",xlab="x",ylab="term33(x,t)",
          xlim=c(xl,xu),lty=1,main="term33(x,t)",lwd=2,
          col="black");
#
```

```
# term34
  matplot(x=x,y=term34,type="l",xlab="x",ylab="term34(x,t)",
          xlim=c(xl,xu),lty=1,main="term34(x,t)",lwd=2,
          col="black");
#
# term35
  matplot(x=x,y=term35,type="l",xlab="x",ylab="term35(x,t)",
          xlim=c(xl,xu),lty=1,main="term35(x,t)",lwd=2,
          col="black");
# pde3t
  matplot(x=x,y=pde3t,type="l",xlab="x",ylab="pde3t(x,t)",
          xlim=c(xl,xu),lty=1,main="pde3t(x,t)",lwd=2,
          col="black");
```

The plots are parametric in t (see Figs. 5.3-1,2). Default scaling of the y (*ordinate*) variable is used in each call to `matplot`.

`par(mfrow=c(2,2))` specifies a 2×2 matrix of plots on each page.

The following coding is for eq. (3.1-4), with some repetition of the preceding discussion for eqs. (3.1-1,2,3) so that the explanation is self contained.

- Matrices for the three RHS terms of eq. (3.1-4) and the LHS t derivative are defined.

```
#
# IL10(x,t)
  term41=matrix(0,nrow=nx,ncol=nout);
  term42=matrix(0,nrow=nx,ncol=nout);
  term43=matrix(0,nrow=nx,ncol=nout);
   pde4t=matrix(0,nrow=nx,ncol=nout);
```

The terms are named with two numbers. The first number designates the PDE. The second number designates the term.

For example, `term41` is the first term in the fourth PDE (eq. (3.1-4)).

- The three RHS terms, `term41` to `term43`, and the LHS term, `pde4t` are computed over x with an inner `for` and index `ix`, and over t with an outer `for` and index `it`.

```
  for(it in 1:nout){
  for(ix in 1:nx){
  fc1=k5*M2[ix,it]*c1/(c1+IL10[ix,it]);
  term41[ix,it]=DIL10*IL10xx[ix,it];
  term42[ix,it]=fc1;
```

```
    term43[ix,it]=-dIL10*IL10[ix,it];
     pde4t[ix,it]=term41[ix,it]+term42[ix,it]+term43[ix,it];
    }
    }
```

- The three RHS terms and LHS *t* derivative are plotted in 2D with the R utility `matplot`.

```
#
# Plot IL10(x,t) RHS, LHS terms
#
# term41
  par(mfrow=c(2,2));
  matplot(x=x,y=term41,type="l",xlab="x",ylab="term41(x,t)",
          xlim=c(xl,xu),lty=1,main="term41(x,t)",lwd=2,
          col="black");
#
# term42
  matplot(x=x,y=term42,type="l",xlab="x",ylab="term42(x,t)",
          xlim=c(xl,xu),lty=1,main="term42(x,t)",lwd=2,
          col="black");
#
# term43
  matplot(x=x,y=term43,type="l",xlab="x",ylab="term43(x,t)",
          xlim=c(xl,xu),lty=1,main="term43(x,t)",lwd=2,
          col="black");
# pde4t
  matplot(x=x,y=pde4t,type="l",xlab="x",ylab="pde4t(x,t)",
          xlim=c(xl,xu),lty=1,main="pde4t(x,t)",lwd=2,
          col="black");
```

The plot is parametric in *t* (see Fig. 5.4-1). Default scaling of the *y* (*ordinate*) variable is used in each call to `matplot`.
`par(mfrow=c(2,2))` specifies a 2×2 matrix of plots on each page.

The following coding is for eq. (3.1-5), with some repetition of the preceding discussion for eqs. (3.1-1,2,3,4) so that the explanation is self contained.

- Matrices for the three RHS terms of eq. (3.1-5) and the LHS *t* derivative are defined.

```
#
# Ta(x,t)
```

```
term51=matrix(0,nrow=nx,ncol=nout);
term52=matrix(0,nrow=nx,ncol=nout);
term53=matrix(0,nrow=nx,ncol=nout);
pde5t=matrix(0,nrow=nx,ncol=nout);
```

The terms are named with two numbers. The first number designates the PDE. The second number designates the term.

For example, term51 is the first term in the fifth PDE (eq. (3.1-5)).

• The three RHS terms, term51 to term53, and the LHS term, pde5t are computed over x with an inner for and index ix, and over t with an outer for and index it.

```
for(it in 1:nout){
for(ix in 1:nx){
fc=c/(c+IL10[ix,it]);
term51[ix,it]=DTa*Taxx[ix,it];
term52[ix,it]=(k6*M1[ix,it]+lam*Mc[ix])*fc;
term53[ix,it]=-dTa*Ta[ix,it];
pde5t[ix,it]=term51[ix,it]+term52[ix,it]+term53[ix,it];
}
}
```

• The three RHS terms and LHS t derivative are plotted in 2D with the R utility matplot.

```
#
# Plot Ta(x,t) RHS, LHS terms
#
# term51
  par(mfrow=c(2,2));
  matplot(x=x,y=term51,type="l",xlab="x",ylab="term51(x,t)",
          xlim=c(xl,xu),lty=1,main="term51(x,t)",lwd=2,
          col="black");
#
# term52
  matplot(x=x,y=term52,type="l",xlab="x",ylab="term52(x,t)",
          xlim=c(xl,xu),lty=1,main="term52(x,t)",lwd=2,
          col="black");
#
# term53
  matplot(x=x,y=term53,type="l",xlab="x",ylab="term53(x,t)",
          xlim=c(xl,xu),lty=1,main="term53(x,t)",lwd=2,
```

```
            col="black");
#
# pde5t
  matplot(x=x,y=pde5t,type="l",xlab="x",ylab="pde5t(x,t)",
          xlim=c(xl,xu),lty=1,main="pde5t(x,t)",lwd=2,
          col="black");
```

The plot is parametric in t (see Fig. 5.5-1). Default scaling of the y (*ordinate*) variable is used in each call to matplot.

par(mfrow=c(2,2)) specifies a 2×2 matrix of plots on each page.

The following coding is for eq. (3.1-6), with some repetition of the preceding discussion for eqs. (3.1-1,2,3,4,5) so that the explanation is self contained.

*   Matrices for the three RHS terms of eq. (3.1-6) and the LHS t derivative are defined.

```
#
# IL1(x,t)
  term61=matrix(0,nrow=nx,ncol=nout);
  term62=matrix(0,nrow=nx,ncol=nout);
  term63=matrix(0,nrow=nx,ncol=nout);
  pde6t=matrix(0,nrow=nx,ncol=nout);
```

The terms are named with two numbers. The first number designates the PDE. The second number designates the term.

For example, term61 is the first term in the sixth PDE (eq. (3.1-6)).

*   The three RHS terms, term61 to term63, and the LHS term, pde6t are computed over x with an inner for and index ix, and over t with an outer for and index it.

```
  for(it in 1:nout){
  for(ix in 1:nx){
  fc=c/(c+IL10[ix,it]);
  term61[ix,it]=DIL1*IL1xx[ix,it];
  term62[ix,it]=(k7*M1[ix,it]+lam*Mc[ix])*fc;
  term63[ix,it]=-dIL1*IL1[ix,it];
  pde6t[ix,it]=term61[ix,it]+term62[ix,it]+term63[ix,it];
  }
  }
```

*   The three RHS terms and LHS t derivative are plotted in 2D with the R utility matplot.

```
#
# Plot IL1(x,t) RHS, LHS terms
#
# term61
  par(mfrow=c(2,2));
  matplot(x=x,y=term61,type="l",xlab="x",ylab="term61(x,t)",
          xlim=c(xl,xu),lty=1,main="term61(x,t)",lwd=2,
          col="black");
#
# term62
  matplot(x=x,y=term62,type="l",xlab="x",ylab="term62(x,t)",
          xlim=c(xl,xu),lty=1,main="term62(x,t)",lwd=2,
          col="black");
#
# term63
  matplot(x=x,y=term63,type="l",xlab="x",ylab="term63(x,t)",
          xlim=c(xl,xu),lty=1,main="term63(x,t)",lwd=2,
          col="black");
#
# pde6t
  matplot(x=x,y=pde1t,type="l",xlab="x",ylab="pde6t(x,t)",
          xlim=c(xl,xu),lty=1,main="pde6t(x,t)",lwd=2,
          col="black");
```

The plot is parametric in t (see Fig. 5.6-1). Default scaling of the y (*ordinate*) variable is used in each call to `matplot`.

`par(mfrow=c(2,2))` specifies a 2×2 matrix of plots on each page.

5.1.2 ODE/MOL routine

The ODE/MOL routine called by `lsodes` is again pde1b in Listing 3.4. The preceding output is for `ncase=2`.

5.1.3 Graphical output

The graphical output from Listing 5.1 follows.

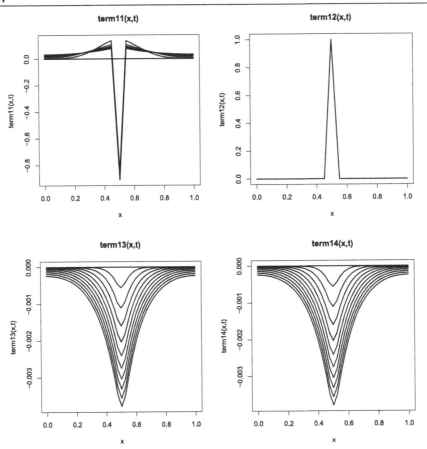

Figure 5.1-1:
term11 to term14, DMun=0.1, ncase=2, 2D

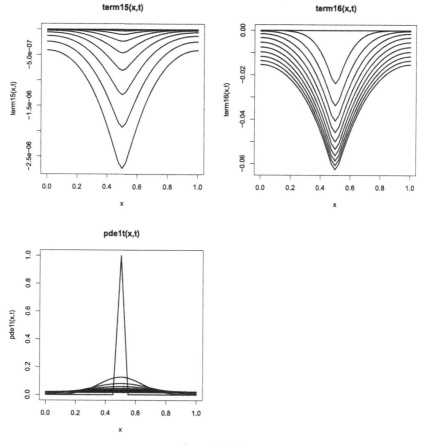

Figure 5.1-2:
term15, term16, pde1t, DMun=0.1, ncase=2, 2D

term12 is the dominant term (through the monocyte rate M[ix]) in determining the diffusion of term11 and the *t* derivative pde1 $= \dfrac{\partial M_{un}(x,t)}{\partial t}$ (which is confirmed in Fig. 4.3-5).

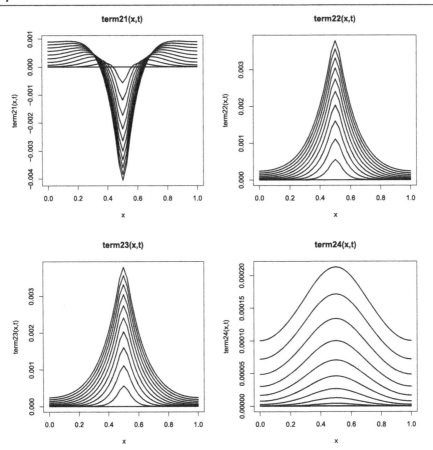

Figure 5.2-1:
term21 to term24, DM1=0.1, ncase=2, 2D

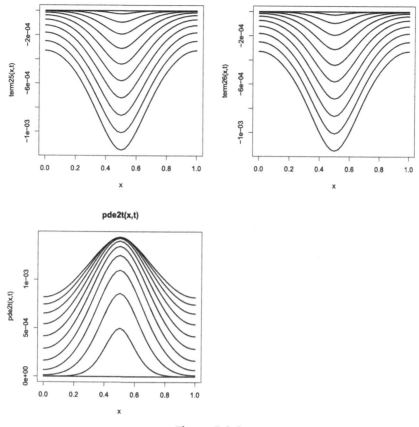

Figure 5.2-2:
term25, term26, pde2t, DM1=0.1, ncase=2, 2D

term21 to term26 are comparable in contributing to the t derivative pde2t $= \dfrac{\partial M_1(x,t)}{\partial t}$ (which is confirmed in Fig. 4.3-5).

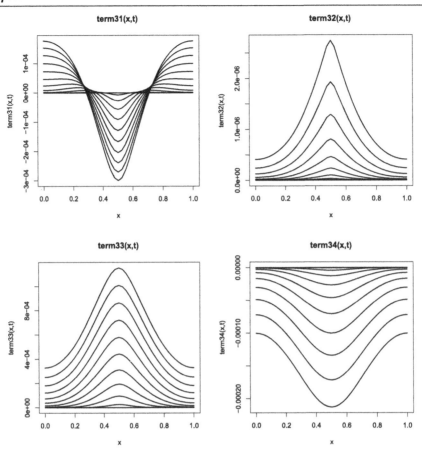

Figure 5.3-1:
term31 to term34, DM2=0.1, ncase=2, 2D

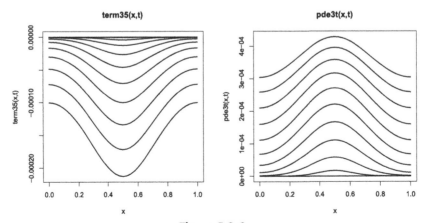

Figure 5.3-2:
term35, pde3t, DM2=0.1, ncase=2, 2D

`term31` to `term35` are comparable in contributing to the t derivative pde3t $\dfrac{\partial M_2(x,t)}{\partial t}$ (which is confirmed in Fig. 4.3-5).

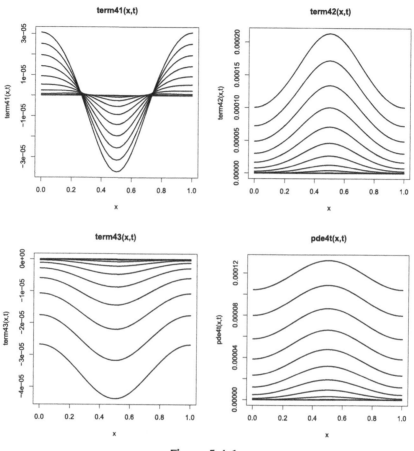

Figure 5.4-1:
`term41` to `term43`, pde4t, DIL10=0.1, ncase=2, 2D

`term41` to `term43` are comparable in contributing to the t derivative pde4t $= \dfrac{\partial IL_{10}(x,t)}{\partial t}$ (which is confirmed in Fig. 4.3-5).

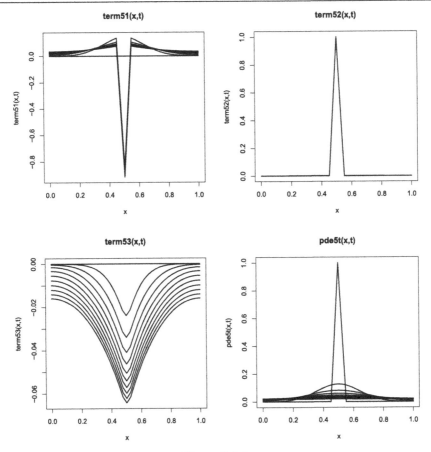

Figure 5.5-1:
term51 to term53, pde5t, DTa=0.1, ncase=2, 2D

term52 is the dominant term (through the myocyte rate Mc[ix]) in determining the diffusion of term51 and the t derivative pde5t $= \dfrac{\partial T_\alpha(x, t)}{\partial t}$ (which is confirmed in Fig. 4.3-6).

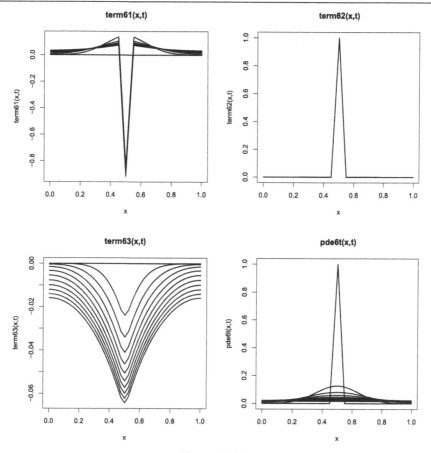

Figure 5.6-1:
term61 to term63, pde6t, DIL1=0.1, ncase=2, 2D

term62 is the dominant term (through the myocyte rate Mc[ix]) in determining the diffusion of term61 and the t derivative pde6t = $\dfrac{\partial IL_1(x,t)}{\partial t}$ (which is confirmed in Fig. 4.3-6).

5.2 Summary and conclusions

The analysis of the RHS terms of eqs. (3.1) in this chapter gives a detailed explanation of the solutions of eqs. (3.1)

$$M_{un}(x,t),\ M_1(x,t),\ M_2(x,t),\ IL_{10}(x,t),\ T_\alpha(x,t),\ IL_1(x,t)$$

$IL_1(x,t)$ is of particular interest since interleuken-1 can cause harm to cardiac tissue. Severe infarction can result in fatal cardiac failure.

References

[1] K. Soetaert, J. Cash, F. Mazzia, Solving Differential Equations in R, Springer-Verlag, Heidelberg, Germany, 2012.

Functions dss004, dss044

A.1 dss004 *listing*

A listing of function dss004 follows.

```
dss004=function(xl,xu,n,u) {
#
# An extensive set of documentation comments detailing
# the derivation of the following fourth order finite
# differences (FDs) is not given here to conserve
# space.  The derivation is detailed in Schiesser,
# W. E., The Numerical Method of Lines Integration
# of Partial Differential Equations, Academic Press,
# San Diego, 1991.
#
# Preallocate arrays
  ux=rep(0,n);
#
# Grid spacing
  dx=(xu-xl)/(n-1);
#
# 1/(12*dx) for subsequent use
  r12dx=1/(12*dx);
#
# ux vector
#
# Boundaries (x=xl,x=xu)
  ux[1]=r12dx*(-25*u[1]+48*u[ 2]-36*u[ 3]+16*u[ 4]-3*u[ 5]);
  ux[n]=r12dx*( 25*u[n]-48*u[n-1]+36*u[n-2]-16*u[n-3]+3*u[n-4]);
#
```

```
# dx in from boundaries (x=xl+dx,x=xu-dx)
  ux[  2]=r12dx*(-3*u[1]-10*u[  2]+18*u[  3]-6*u[  4]+u[  5]);
  ux[n-1]=r12dx*( 3*u[n]+10*u[n-1]-18*u[n-2]+6*u[n-3]-u[n-4]);
#
# Interior points (x=xl+2*dx,...,x=xu-2*dx)
  for(i in 3:(n-2))ux[i]=r12dx*(-u[i+2]+8*u[i+1]-8*u[i-1]+u[i-2]);
#
# All points concluded (x=xl,...,x=xu)
  return(c(ux));
}
```

Listing A.1: dss004

We can note the following details about Listing A.1.

The input arguments are

xl	lower boundary value of x
xu	upper boundary value of x
n	number of points in the grid in x, including the end points
u	dependent variable to be differentiated, an n-vector

The output, ux, is an n-vector of numerical values of the first derivative of u.

The finite difference (FD) approximations are a weighted sum of the dependent variable values. For example, at point i


```
 for(i in 3:(n-2))ux[i]=r12dx*(-u[i+2]+8*u[i+1]-8*u[i-1]+u[i-2]);
```


The weighting coefficients are -1, 8, 0, -8, 1 at points i-2, i-1, i, i+1, i+2, respectively. These weighting coefficients are antisymmetic (opposite sign) around the center point i because the computed first derivative is of odd order. If the derivative is of even order, the weighting coefficients would be symmetric (same sign) around the center point.

For i=1. the dependent variable at points i=1,2,3,4,5 is used in the FD approximation for ux[1] to remain within the x domain (fictitious points outside the x domain are not used).

```
ux[1]=r12dx*(-25*u[1]+48*u[2]-36*u[3]+16*u[4]-3*u[5]);
```

Similarly, for i=2, points i=1,2,3,4,5 are used in the FD approximation for ux[2] to remain within the x domain (fictitious points outside the x domain are avoided).

```
ux[2]=r12dx*(-3*u[1]-10*u[2]+18*u[3]-6*u[4]+u[5]);
```

At the right boundary $x = x_u$, points at i=n,n-1,n-2,n-3,n-4 are used for ux[n],ux[n-1] to avoid points outside the x domain.

In all cases, the FD approximations are fourth order correct in x.

A.2 dss044 *listing*

A listing of function dss044 follows.

```
  dss044=function(xl,xu,n,u,ux,nl,nu) {
#
# The derivation of the finite difference
# approximations for a second derivative are
# in Schiesser, W. E., The Numerical Method
# of Lines Integration of Partial Differential
# Equations, Academic Press, San Diego, 1991.
#
# Preallocate arrays
  uxx=rep(0,n);
#
# Grid spacing
  dx=(xu-xl)/(n-1);
#
# 1/(12*dx**2) for subsequent use
  r12dxs=1/(12*dx^2);
#
# uxx vector
#
# Boundaries (x=xl,x=xu)
  if(nl==1)
    uxx[1]=r12dxs*
```

```
                 (45*u[  1]-154*u[  2]+214*u[  3]-
                 156*u[  4] +61*u[  5] -10*u[  6]);
  if(nu==1)
    uxx[n]=r12dxs*
                 (45*u[  n]-154*u[n-1]+214*u[n-2]-
                 156*u[n-3] +61*u[n-4] -10*u[n-5]);
  if(nl==2)
    uxx[1]=r12dxs*
                 (-415/6*u[  1] +96*u[  2]-36*u[  3]+
                    32/3*u[  4]-3/2*u[  5]-50*ux[1]*dx);
  if(nu==2)
    uxx[n]=r12dxs*
                 (-415/6*u[  n] +96*u[n-1]-36*u[n-2]+
                    32/3*u[n-3]-3/2*u[n-4]+50*ux[n]*dx);
#
# dx in from boundaries (x=xl+dx,x=xu-dx)
    uxx[  2]=r12dxs*
                 (10*u[  1]-15*u[  2]-4*u[  3]+
                 14*u[  4]- 6*u[  5]  +u[  6]);
    uxx[n-1]=r12dxs*
                 (10*u[  n]-15*u[n-1]-4*u[n-2]+
                 14*u[n-3]- 6*u[n-4]  +u[n-5]);
#
# Remaining interior points (x=xl+2*dx,...,
# x=xu-2*dx)
  for(i in 3:(n-2))
    uxx[i]=r12dxs*
                 (-u[i-2]+16*u[i-1]-30*u[i]+
                 16*u[i+1]    -u[i+2]);
#
# All points concluded (x=xl,...,x=xu)
  return(c(uxx));
}
```

Listing A.2: dss044

We can note the following details about Listing A.2.

The input arguments are

xl	lower boundary value of *x*
xu	upper boundary value of *x*
n	number of points in the grid in *x*, including the end points
u	dependent variable to be differentiated, an *n*-vector
ux	first derivative of u with boundary condition (BC) values, an *n*-vector
nl	type of boundary condition at x=xl
	1: Dirichlet BC
	2: Neumann BC
nu	type of boundary condition at x=xu
	1: Dirichlet BC
	2: Neumann BC

The output, uxx, is an *n*-vector of numerical values of the second derivative of u.

The finite difference (FD) approximations are a weighted sum of the dependent variable values. For example, at point i

```
for(i in 3:(n-2))
  uxx[i]=r12dxs*
        (-u[i-2]+16*u[i-1]-30*u[i]+
     16*u[i+1]    -u[i+2]);
```

The weighting coefficients are -1, 16, -30, 16, -1 at points i-2, i-1, i, i+1, i+2, respectively. These weighting coefficients are symmetric around the center point i because the computed second derivative is of even order. If the derivative is of odd order, the weighting coefficients would be antisymmetric (opposite sign) around the center point.

For nl=2 and/or nu=2 the boundary values of the first derivative are included in the FD approximation for the second derivative, uxx. For example, at x=xl (with nl=2),

```
if(nl==2)
  uxx[1]=r12dxs*
        (-415/6*u[  1] +96*u[  2]-36*u[  3]+
      32/3*u[  4]-3/2*u[  5]-50*ux[1]*dx);
```

In computing the second derivative at the left boundary, uxx[1], the first derivative at the left boundary is included, that is, ux[1]. In this way, a Neumann BC is accommodated (ux[1] is included in the input argument ux).

For nl=1, only values of the dependent variable (and not the first derivative) are included in the weighted sum.

```
if(nl==1)
   uxx[1]=r12dxs*
           (45*u[  1]-154*u[  2]+214*u[  3]-
           156*u[  4] +61*u[  5] -10*u[  6]);
```

The dependent variable at points i=1,2,3,4,5,6 is used in the FD approximation for uxx[1] to remain within the x domain (fictitious points outside the x domain are not used).

Six points are used rather than five (as in the centered approximation for uxx[i]) since the FD applies at the left boundary and is not centered (around i). Six points provide a fourth order FD approximation which is the same order as the FDs at the interior points in x.

Similar considerations apply at the upper boundary value of x with nu=1,2.

Robin boundary conditions can also be accommodated with nl=2, nu=2. In all three cases, Dirichlet, Neumann and Robin, the boundary conditions can be linear and/or nonlinear.

Index